T0321275

# GEOMATERIALS
## UNDER THE MICROSCOPE

### A COLOUR GUIDE

building stone, roofing slate, aggregate,
concrete, mortar, plaster, bricks,
ceramics, and bituminous mixtures

## Jeremy P. Ingham

BSc (Hons), MSc, DipRMS, CEng, MInstNDT, EurGeol, CGeol, CSci, FGS, FRGS, MIAQP

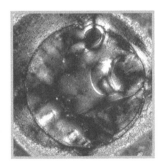

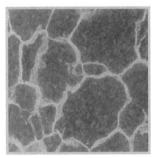

MANSON
PUBLISHING

Copyright © 2011 Manson Publishing Ltd

ISBN: 978-1-84076-132-0

A CIP catalogue record for this book is available from the British Library.

For full details of all Manson Publishing titles please write to:
Manson Publishing Ltd, 73 Corringham Road, London NW11 7DL, UK.
Tel: +44(0)20 8905 5150
Fax: +44(0)20 8201 9233
Website: www.mansonpublishing.com

Commissioning editor: Jill Northcott
Project manager: Kate Nardoni
Copy editor: Joanna Brocklesby
Book design and layout: Cathy Martin
Colour reproduction: Tenon & Polert Colour Scanning Ltd, Hong Kong
Printed by: Butler, Tanner & Dennis, Frome, UK

# CONTENTS

**Preface** 5

**Acknowledgements** 5

**Author profile** 5

**Abbreviations** 6


**Chapter 1**
  **Introduction** 7

Overview of geomaterials and petrography 7

Petrographic techniques 10

Sampling and sample preparation 14


**Chapter 2**
  **Building stone** 21

Introduction 21

Testing building stone 22

Petrographic examination and complementary
  techniques 24

Stone from igneous rocks 24

Stone from sedimentary rocks 32

Stone from metamorphic rocks 44

Petrography of stone defects and decay 47


**Chapter 3**
  **Roofing slate** 51

Introduction 51

Testing roofing slates 51

Petrographic examination and complementary
  techniques 52

Properties of roofing slates 52

Weathering and deterioration of roofing slate 56

Interpretation of slate test results 58


**Chapter 4**
  **Aggregates** 61

Introduction 61

Petrographic examination and complementary
  techniques 61

Aggregate type 61

Aggregate grading, shape, and surface texture 64

Soundness, impurities, and undesirable
  constituents of aggregates 66

Potential alkali-reactivity of aggregate for
  concrete 69

Matching aggregates for conservation 74


**Chapter 5**
  **Concrete** 75

Introduction 75

Assessment of concrete structures 75

Petrographic examination and complementary
  techniques 76

Investigating the composition and quality of
  concrete 76

Examining deteriorated and damaged concrete 96


**Chapter 6**
  **Concrete products** 121

Introduction 121

Petrographic examination 122

Architectural cast stone 122

Aircrete products 124

Calcium silicate units 124

Asbestos cement products 126

**Chapter 7**
  **Floor finishes**      129

Introduction      129

Petrographic examination      130

Floor screed      130

Terrazzo      132

Synthetic resin floor covering      133

**Chapter 8**
  **Mortar, plaster, and render**      137

Overview      137

Gypsum-based      139

Lime-based      143

Portland cement-based      152

Specialist mortars      159

**Chapter 9**
  **Bricks, terracotta, and other ceramics**      163

Introduction      163

Petrographic examination and complementary
  techniques      163

Clay brick      164

Terracotta      166

Ceramic tile      168

Architectural glass      170

**Chapter 10**
  **Bituminous mixtures**      171

Introduction      171

Petrographic examination      172

Examples of bituminous mixtures      172

**Appendix A**      175

Suppliers of petrographic equipment, training,
  and literature      175

**Appendix B**      176

Staining techniques for geomaterials
  petrography      176

Etching and staining techniques for cement
  minerals and slags      178

**References and further reading**      180

**Index**      189

# PREFACE

When I embarked upon my career as a construction materials petrographer I found that there were no specific books available to help me learn the trade. Instead, like many others my knowledge was gained directly from an experienced expert who guided and supervised my early work. I have since been fortunate enough to have enjoyed a broad range of industrial and research experience that has enabled me to compile this guidebook, the process of which has taken 12 years.

By providing the first comprehensive guide to the petrography of geomaterials I have attempted to make the specialized knowledge of the petrographer available to all. It is intended not only for microscopists but for anyone with an interest in modern and historic construction materials from the fields of geology, architecture, surveying, engineering, construction, archaeology, conservation, materials science, and forensic science. The format of the book is a combination of an atlas of high-quality colour photomicrographs of geomaterials, along with explanations of their petrographic properties and how to interpret them. An introductory chapter provides an overview of geomaterials practice and a state-of-the-art review of petrographic techniques. Each of the subsequent chapters covers a different group of construction materials and includes an explanation of their history, manufacture, and use in construction. The length of each of the individual materials chapters reflects the relative importance of the material as an area of commercial petrographic study.

As a discipline, construction materials petrography is relatively young and dynamic. I intend to update this book at regular intervals to reflect the growth in knowledge and changes to the state of practice. I would welcome any contributions and case study examples to be considered for inclusion in subsequent editions.

# ACKNOWLEDGEMENTS

I wish to express my gratitude to Barry Hunt of IBIS Ltd, Mike Eden of Geomaterials Research Services Ltd, and Kate McHardy of Oxford Cryosystems Ltd for generously contributing photomicrographs for inclusion in this book. I acknowledge with thanks the following companies who gave permission for me to use photographs of their products: Concrete Experts International ApS, Leica Microsystems GmbH, and Logitech Ltd.

# AUTHOR PROFILE

Jeremy Ingham is a Consulting Geomaterials Engineer who provides consultancy and investigation services for construction projects worldwide, involving materials technology, forensic engineering, and asset management planning. Commissions range from preconstruction advice, quality control during construction, condition surveys and remediation of existing structures, failure investigations, and expert witness services.

As a student he gained a BTEC HND in Geological Technology, a BSc in Geology, and an MSc in Geomaterials. His experience of construction materials petrography started with 3 years of working as a geological laboratory technician, which gave him a thorough grounding in the practical aspects of petrography. Next followed a decade spent working as a petrographer, investigating a wide range of modern and historic construction materials. During this time he was responsible for commissioning and managing petrographic facilities and training staff in petrographic methods. Since then he has acted as a consultant, both for new construction projects and conservation of historic buildings. An accomplished microscopist, he is both a Fellow of the Royal Microscopical Society and holder of the Diploma of the Royal Microscopical Society. He is Chairman of the Concrete Society's Petrography Working Group and a member of the Geological Society's Applied Petrography Group.

Contact: www.jeremyingham.com

# ABBREVIATIONS

| | |
|---|---|
| AAC | autoclaved aerated concrete |
| AAR | alkali–aggregate reaction |
| AAS | atomic absorption spectroscopy |
| ACR | alkali–carbonate reaction |
| ASR | alkali–silica reaction |
| ASTM | American standards |
| CEN | Comité Européen de Normalisation |
| CSF | condensed silica fume |
| C-S-H | calcium silicate hydrates |
| DEF | delayed ettringite formation |
| DSC | differential scanning calorimetry |
| DTA | differential thermal analysis |
| EDS | energy dispersive X-ray spectroscopy |
| EN | European standards |
| EPM | electron probe microanalysis |
| ESEM | environmental scanning electron microscopy |
| FESEM | field emission scanning electron microscopy |
| FGD | flue gas desulfurization |
| GGBS | ground granulated blastfurnace slag |
| HAC | high-alumina cement |
| HF | hydrofluoric acid |
| ICP-AES | inductively coupled plasma atomic emission spectroscopy |
| KOH | potassium hydroxide |
| NR (in captions) | normal reflected light |
| PCD | popcorn calcite deposition |
| PFA | pulverized fuel ash |
| PPR (in captions) | plane-polarized reflected light |
| PPT (in captions) | plane-polarized transmitted light |
| PVA | polyvinyl acetate |
| PVC | polyvinyl chloride |
| PVDC | polyvinylidene dichloride |
| RAAC | reinforced autoclaved aerated concrete |
| SBR | styrene butadiene rubber |
| SEM | scanning electron microscopy |
| SMA | stone mastic asphalt |
| SRPC | sulfate-resisting Portland cement |
| TGA | thermogravimetric analysis |
| TSA | thaumasite form of sulfate attack |
| UV (in captions) | reflected fluorescent light |
| W/C | water/cement ratio |
| XPR (in captions) | cross-polarized reflected light |
| XPT (in captions) | cross-polarized transmitted light |
| XRD | X-ray diffraction |
| XRF | X-ray fluorescence |

# Introduction

## OVERVIEW OF GEOMATERIALS AND PETROGRAPHY

Geomaterials are defined as 'processed or unprocessed soils, rocks or minerals used in the construction of buildings or structures, including man-made construction materials manufactured from soils, rocks or minerals' (Fookes, 1991). The definition deliberately includes man-made materials such as bricks or cement but excludes allied engineering materials whose manufacturing is more extensive, such as steel and synthetic paints.

Geologically derived materials have been used in construction for much of human history and, being geologically common, are available worldwide. Strong rocks provide dimension stone for masonry, cladding, and flooring, while thinly bedded or cleavable rocks are suitable for roofing. Limestone can be calcined to produce lime or cement, which can be mixed with sand and water to make mortar for bonding masonry units. Gypsum is heated to produce plaster and plasterboard. Large stone blocks are used as armourstone protection for coastlines and riverbanks. Aggregates from crushed rock are used for roadstone, either loose or in layers bound by bitumen, as macadam and asphalt. Both crushed rock and natural gravel aggregates are mixed with Portland cement and water to produce concrete. Aggregates of all origins are used as fill materials for earthworks. Pure silica sands are melted and cast into glass for windows and other architectural elements. Clay soils can be used to make sun-dried or kiln-fired bricks for masonry construction, architectural terracotta, and ceramic tiles. Figure 1 provides a summary of the main resources used for bulk construction, showing the links between the various raw materials, the finished products, and their uses.

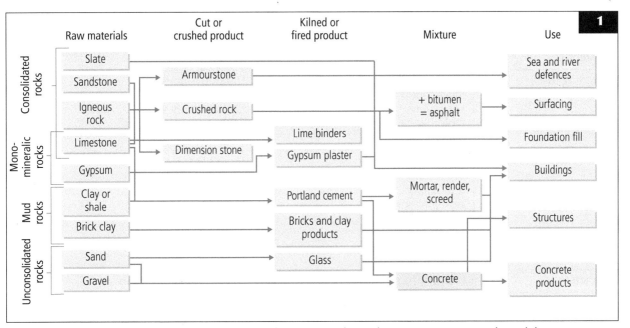

**1** Flow chart showing the relationships between the main geological construction materials and their uses (adapted from Woodcock, 1994).

The economic importance of geomaterials and their contribution to the global construction industry cannot be overestimated. For example, the ex-works value of raw materials production for the construction industry in Britain alone was estimated as being £2,105,000,000 in 2004 (Taylor *et al.*, 2006). *Table 1* lists the quantities of different geomaterials produced for construction in the British Isles.

Testing of construction materials to ensure that they are fit for purpose has been an essential part of the construction process since the early 19th century. The second half of the 19th century saw the emergence of independent materials testing companies (Doran & Cockerton, 2006). Since the 1920s there has been a growing cooperation between civil engineers and geologists to develop the knowledge of geomaterials in the construction industry. During the last 40 years, geomaterials practice has evolved into a discipline in its own right. Today, geomaterials practitioners apply geological and materials engineering knowledge to construction and civil engineering projects, ranging from new building and infrastructure developments through to monitoring, maintenance, refurbishment, and conservation of existing structures. They are involved in the full lifecycle of built assets with tasks typically comprising:

- Location of new construction material resources.
- Evaluation of materials proposed for use in construction.
- Routine quality control testing and monitoring during construction.
- Investigation of deterioration during service.
- Screening existing structures for potential problems.
- Monitoring repair and restoration programmes.
- Expert witness service.

Clients of the geomaterials engineer may include architects, engineers, specifiers, surveyors, commercial property owners/managers, domestic property owners, manufacturers, contractors, and government agencies. Petrographic examination is one of the most powerful investigative tools available to the geomaterials engineer. It involves using the polarizing microscope to examine thin sections or polished surfaces of samples, in the same way that geologists examine rock samples. This may be supplemented by various other microscopical techniques, chemical analysis, and simple physical tests. First used for examination of rock in 1849, petrography was soon applied

**Table 1** Production of construction geomaterials in the United Kingdom, 2004 (Taylor *et al.*, 2006)

| Material | Type | Quantity |
|---|---|---|
| Building stone | Igneous rock<br>Sandstone<br>Limestone<br>Dolomite<br>Slate | 189,000 tonnes<br>439,000 tonnes<br>226,000 tonnes<br>8,000 tonnes<br>763,000 tonnes |
| Aggregate | Concrete aggregate<br>Roadstone<br>Rail ballast<br>Armourstone<br>Construction fill<br>For glass making | 81,293,000 tonnes<br>47,768,000 tonnes<br>3,630,000 tonnes<br>333,000 tonnes<br>39,313,000 tonnes<br>5,011,000 tonnes |
| Cement | Finished product | 11,405,000 tonnes |
| Concrete | Ready-mixed | 22,856,000 cubic metres |
| Brick | Clay brick | 2,707,000,000 bricks |
| Gypsum | For plaster products | 1,686,000 tonnes |

to other construction materials. The technique requires specialist equipment and operators with appropriate qualifications and petrographic experience. Through the microscope the petrographer can determine the composition of geomaterials, assess their quality, and investigate the causes and extent of deterioration. Photographs taken through the microscope (photomicrographs) provide compelling evidence for clients. Around the world, tens of thousands of petrographic examinations are performed on construction materials each year. The main requirements for a commercial geomaterials petrography laboratory are:

- It must led by an experienced petrographer/geomaterials practitioner.

- The examinations should be conducted using very high-quality optical microscopes.
- It should have good in-house specimen preparation facilities.
- An extensive collection of petrographic specimens should be available for the petrographers' reference.
- There should be access to complementary microscopical and chemical analysis techniques.

Commercially, petrographic examination is conducted in accordance with the published standards listed in *Table 2*. Standard methods have still to be agreed for certain types of material and in certain geographic regions.

**Table 2** Current standards for the petrographic examination of geomaterials

| Material | British standard | European standard | American standard |
|---|---|---|---|
| Building stone | Use European standard | EN 12407 [1] | WK2609 [2] |
| Roofing slate | Use European standard | EN 12326-2 [3] | None* |
| Rock | None, use ISRM 1977 [4] and BS 5930 [5] | | |
| Armourstone | Use European standard | Use EN 932-3 [6] | None* |
| Aggregate | BS 812-104 [7] | EN 932-3 [6] | ASTM C295 [8] |
| Concrete | None – use American standard | None – use American standard | ASTM C856 [9] |
| Mortar | None – use American standard | None – use American standard | ASTM C1324 [10] |
| Bricks and ceramics | None* | None* | None* |
| Bituminous mixtures | None* | None* | None* |

\* Adapt the techniques described in other petrography standards as appropriate.
1  British Standards Institution (2007). *Natural Stone Test Methods – Petrographic Examination: BS EN 12407.* BSI, London.
2  ASTM International (2006). *New Standard Guide for Petrographic Examination of Dimension Stone.* WK2609 (proposed new standard in draft form and under development). ASTM International, Philadelphia.
3  British Standards Institution (2000). *Slate and Stone Products for Discontinuous Roofing and Cladding – Part 2: Methods of Test: BS EN 12326-2.* BSI, London.
4  International Society for Rock Mechanics (ISRM) (1977). *Rock Characterisation, Testing and Monitoring. Suggested Method for the Petrographic Description of Rock.* Pergamon Press, Oxford.
5  British Standards Institution (1999). *Code of Practice for Site Investigations: BS 5930.* BSI, London.
6  British Standards Institution (1997). *Tests for General Properties of Aggregates – Part 3: Procedure and Terminology for Simplified Petrographic Description: BS EN 932-3.* BSI, London.
7  British Standards Institution (1994). *Testing Aggregates – Part 104: Method for Qualitative and Quantitative Petrographic Examination of Aggregates: BS 812.* BSI, London.
8  ASTM International (2008). *Standard Guide for Petrographic Examination of Aggregates for Concrete. ASTM C295-08.* ASTM International, Philadelphia.
9  ASTM International (2004). *Standard Practice for the Petrographic Examination of Hardened Concrete. ASTM C856-04.* ASTM International, Philadelphia.
10 ASTM International (2005). *Standard Test Method for Examination and Analysis of Hardened Masonry Mortar. ASTM C1324-05.* ASTM International, Philadelphia.

Where a specific standard is lacking, it is common practice for laboratories offering petrographic services to work in accordance with an in-house 'test procedure', which adapts accepted standards from the nearest material in terms of similarity.

## PETROGRAPHIC TECHNIQUES

### VISUAL AND LIGHT MICROSCOPICAL EXAMINATION

Light microscopy (also called optical microscopy) is at the heart of petrography and, combined with visual examination, it forms the basis of commercial petrographic examinations. The main suppliers of high-quality light microscopes and their accessories are listed in Appendix A. If required, light microscopy can be supplemented with a range of complementary techniques to achieve the objectives of the petrographic examination. These are discussed in Complementary techniques, p. 13.

Following arrival in the laboratory, samples of geomaterials are first examined in the as-received condition using the unaided eye. At this stage a few simple physical and chemical tests may be performed on the hand specimen, such as scratch testing to assess relative hardness (Moh's scale), or acid drop testing (with dilute hydrochloric acid) to determine if carbonate minerals are present. Colour is usually assessed visually in the hand specimen by comparison with standard colour charts such as the Munsell rock colour chart (Geological Society of America, 1991) or the Munsell soil colour chart (Munsell, 1994). The hand specimen is then examined using a low-power stereo-zoom microscope (2) at magnifications of typically up to ×50. In certain circumstances a slice of the sample may be finely ground on one surface to aid the low-power microscopical examination. In addition to being an observational tool, the visual and low-power microscopical examinations are used to determine the most appropriate location for thin section specimens and/or highly polished specimens to be taken for further, more detailed high-power microscopical examination. Specimen preparation is discussed in Sampling and sample preparation, p. 15.

High-power microscopical examination is conducted using a polarizing microscope capable of magnifications up to ×600 (3). Thin section specimens are examined in plane-polarized or cross-polarized transmitted light.

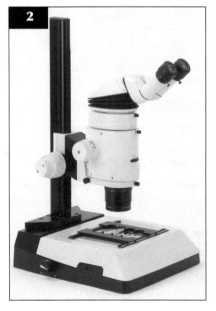

**2** A stereo-zoom low-power microscope for examining samples in hand specimen. (Courtesy of Leica Microsystems UK Ltd.)

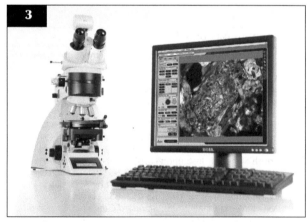

**3** A polarizing microscope set up for combined transmitted light, reflected light, and fluorescence microscopy of thin sections and highly polished specimens. With attached digital image capture system. (Courtesy of Leica Microsystems UK Ltd.)

In addition, various compensator plates may be inserted into the light path to determine different optical properties of minerals. The most commonly used compensator is the gypsum plate (sensitive tint plate) which is used for determining the directions of fast and slow rays in crystals, which is particularly useful for identifying feldspars. Figures **4–6** provide a comparison showing the same view of a rock sample in plane-polarized light, cross-polarized light, and cross-polarized light with the gypsum plate inserted. Fluorescent dyes may be added to the consolidating resin during sample preparation to aid the examination of cracks and pore structures when the specimen is viewed in combination with a strong light source and an excitation filter.

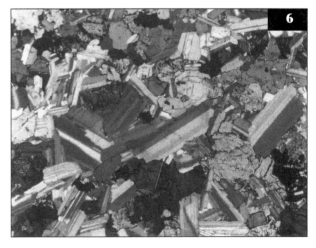

**4–6** Thin section specimen of dolerite viewed in plane-polarized transmitted light (**4**), cross-polarized transmitted light (**5**), and in cross-polarized transmitted light with the gypsum plate inserted into the light path (**6**); ×35.

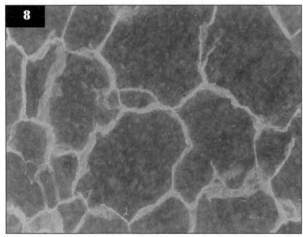

This is termed 'fluorescence microscopy' and Figures **7** and **8** show a comparison between a rock sample in cross-polarized light and fluorescent light, illustrating how the fluorescence view defines porous areas. Sometimes highly polished specimens are prepared for examination in reflected light (brightfield and darkfield). Reflected light is particularly useful for examination of opaque minerals and cements. Figures **9** and **10** show a comparison of ore minerals shown in transmitted light and reflected light. Each of the plate captions in this book includes an abbreviation to show the viewing light of the photomicrograph as follows:

| Applied light | Abbreviation |
| --- | --- |
| Plane-polarized transmitted | PPT |
| Cross-polarized transmitted | XPT |
| Plane-polarized reflected | PPR |
| Cross-polarized reflected | XPR |
| Normal reflected | NR |
| Reflected fluorescence | UV |

Each photomicrograph caption also includes the viewing magnification which correspond to the following fields of view:

| Magnification | Approximate field of view (mm) |
| --- | --- |
| ×35 | 4.5 |
| ×75 | 2.5 |
| ×100 | 1.8 |
| ×150 | 1.0 |
| ×200 | 0.8 |
| ×300 | 0.5 |
| ×600 | 0.2 |

**7, 8** Fluorescent yellow resin impregnated thin-section specimen of marble. Viewed in cross-polarized transmitted light (**7**) and in reflected fluorescent light (**8**); ×150.

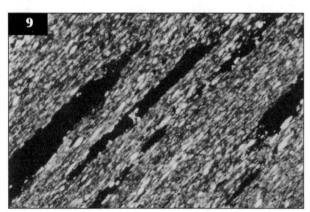

**9** Roofing slate in thin section viewed in plane-polarized transmitted light. Inclusions of pyrite appear opaque (black); PPT, ×75.

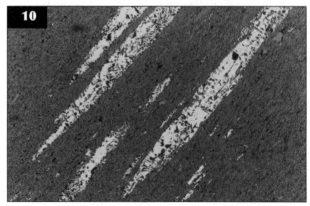

**10** Highly polished specimen of roofing slate viewed in reflected light. Inclusions of pyrite reflect light (yellowish-white); XPR, ×75.

## PHOTOMICROGRAPHY AND IMAGE ANALYSIS

Photographic images can be taken through the microscope (photomicrography) by attaching a camera on to the vertical tube of the microscope's trinocular head. Until recently 35 mm film cameras were used as standard for photomicrography. While these provided high-quality images in the form of photographic prints and slides, they had the disadvantage of having to wait for the film to be processed before one could view and utilize the images. Recent developments in digital image recording have brought rapid changes to photography and, in consequence, also to photomicrography. Film cameras have now been replaced by digital cameras that provide instantly available images, which can be readily manipulated and enhanced using computer software, to suit the required purpose. A basic guide to the practical aspects of digital light micrography is provided by Entwistle (2003b & 2004).

As a result of the move towards digital information, petrographic examination reports are now sent to clients in digital format using electronic mail, in addition to the traditional paper copy sent by conventional mail services. Digital technology has reduced the report production and delivery times for petrographic investigations. Having photomicrographs in digital form also makes it easy to put them up on internet websites, either for public view, or for clients' private use in a password-protected area (Entwistle, 2002).

Image analysis is being increasingly used to measure various properties of construction materials through the microscope. Measurements are obtained by analysis of digital photomicrographs using computer software, which can be highly successful. However, it should be noted that image analysis techniques are only accurate if sufficient image contrast is present to allow accurate identification of the subject being measured. Special sample preparation may be required to ensure this. Also, it must be remembered that the results are only valid if the plane of section analysed is truly representative of the sample. Applications for image analysis of construction materials currently under development include:

- Modal analysis of rock and building stone for mineralogical composition.
- Determination of textures within rock and building stone.
- Determination of the shape, size, and particle size distribution (grading) of aggregates.
- Modal analysis of fine aggregate for mineralogical composition.
- Modal analysis of concrete for aggregate content, aggregate particle shape, aggregate grading, cement content, and air void content.
- Measurement of water/cement ratio of concrete.
- Modal analysis of mortars, renders, and screeds for aggregate content, aggregate particle shape, aggregate grading, binder content, and air void content.
- Assessment of fire damage to concrete by quantifying heat-induced colour changes.

## Complementary techniques

A high-quality petrological microscope can usefully magnify up to 600 times with a maximum resolution of around 1 μm (microns). Where closer examination is required scanning electron microscopy (SEM) is invaluable as it achieves significantly higher magnifications of up to 50,000 times and resolutions of around 7 nm (nanometres), by using a beam of high-energy electrons to replace the light of conventional microscopes. In addition to their high-resolution capabilities, electron microscopes have great depth of field, producing electron images that have a three-dimensional effect. The on-board electron probe microanalysis system (EPM) allows inorganic bulk elemental chemical analysis of characteristic X-ray spectra emitted by samples. The EMP has two modes of operation of importance to petrographers. 'Spot analysis' focuses on selected points of the sample such as individual crystals to give a semiquantitative elemental analysis, while 'elemental mapping' scans along a square line raster over a larger area of the sample surface (up to 30 mm × 15 mm) to allow observation of variations in element distributions.

Mineralogical analysis by X-ray diffraction (XRD) can be helpful for identification of crystalline minerals and decay reaction products, when the optical properties do not allow definitive microscopical identification. XRD has the advantage of actually identifying the minerals present, although estimation of the relative amounts is at best only semiquantitative. Other spectroscopic analysis methods such as X-ray fluorescence (XRF), atomic absorption (AAS), and atomic emission (ICP-AES) give very precise determinations of elemental composition. The elemental composition data from these techniques require skilled interpretation by the analyst to identify the proportions of mineral phases present. The above techniques will only detect inorganic compounds and, to identify the presence of organic materials, infrared spectroscopy is the method of choice.

## Future developments

Optical microscopes have now entered the digital age and modern microscopes tend to come with a computer attached. The major microscope manufacturers now offer options for automation of their optical microscopes with a range of functions being controlled via a touch-sensitive screen. Motors automatically move the objective, substage condenser, and the image projection system; all of which is optimized quite automatically for quality of illumination, resolution, and even focus. Combined with digital image capture systems and image analysis software we now have a real possibility of automated sample examination. However, an experienced petrographer will always be required to supervise the process and check the results. Automation offers huge improvements in productivity and potentially significant reductions in the unit price of examinations. Automated modal analysis is likely to become a serious alternative for a number of standard tests that have traditionally been performed by cheaper (and arguably less accurate) chemical analysis methods. For example, the mix proportions of hardened concrete and mortars could routinely be determined by modal analysis instead of chemical analysis.

Digital technology is changing the way that the findings of petrographic examinations are presented. Digital image capture allows a series of sequential micrograph frames to be animated into movies to illustrate talks and websites (Entwistle, 2003a). Digital movies demonstrating petrographic features of samples could be included within the report submitted to the client.

The recent development of the birefringence imaging microscope offers the possibility of interesting applications for examination of geomaterials. The equipment consists of a motorized rotating polarizer that is fitted below the sample stage and a digital camera that fits on the trinocular microscope head, both attached to a computer. As the motorized polarizer rotates, the camera collects birefringence data at a number (between five and fifty) of different positions. The data are then processed by special computer software to produce various types of false colour image. The birefringence microscope excels at detecting strain and defects within materials. Figures 11 and 12 show examples of false colour images obtained using the birefringence microscope. Existing applications include quality control of industrial diamonds, silicon carbide abrasives, and glass, study of decomposition of biomaterials, and mapping of collagen in heart valves. Potential applications for geomaterials include investigation of rock microstructure (including the

investigation of bowing marble panels), detection of alkali–silica reactive aggregates, and identification of flaws in a variety of construction products.

Electron microscopy is one of the main complementary techniques used in conjunction with optical microscopy hence developments in the field of electron microscopy concern the petrographer. The resolving power of the electron microscope is continually improving, with modern field emission scanning electron microscopy (FESEM) now providing magnifications of up to 550,000 times and resolutions down to 0.5 nm. One disadvantage of conventional electron microscopy is that the sample has to be viewed in a vacuum. Recent advances have allowed hydrated samples to be imaged using environmental scanning electron microscopy (ESEM) or alternatively, by soft X-ray transmission microscopy. These methods allow 'live' examination and analysis of geomaterials undergoing reactions, for example, hydrating cement paste or carbonating lime. Also, the latest cryotransfer SEM allows sensitive hydrating specimens to be set in a stable state by quick-freezing, enabling previously impossible examinations. Advances in electron microscopy will continue to improve our understanding of geomaterials and reactions that they undergo.

# SAMPLING AND SAMPLE PREPARATION

## SAMPLING

Samples of construction materials may be obtained during the manufacturing production run or, alternatively, from structures during construction or while they are in service. The objectives of the materials investigation and details of the proposed laboratory testing should be clearly defined, before any sampling is attempted. A coordinated sampling programme should be prepared by persons experienced with the investigation of construction materials and built structures. The number of samples required to achieve the investigation objective will depend on the purpose of the investigation/testing, the size of the structure, the types of construction used, and the number of construction phases.

Sampling schemes usually comprise one of two types (or a combination of both). The first is essentially a random or even spread of samples across a structure or production run, to ascertain representatively the general materials' characteristics and quality. The second scheme is more targeted to address specific issues, such as investigating suspected defects identified by visual survey. In either case, the investigator must clearly understand the degree of sampling bias within the sampling scheme. Ideally the petrographer should be involved in devising the investigation plan and be present during sampling operations.

The actual sampling technique and the type of sample required for petrographic examination depend on the type of material being sampled, the objectives of the investigation, and if any other types of tests are to be

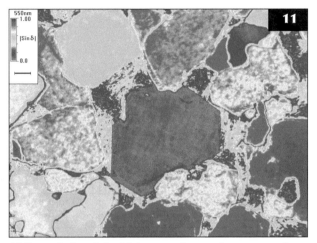

**11** Retardation ($|\sin \delta|$) image of a sandstone sample obtained using the birefringence microscope imaging system. This shows the relative birefringence of the different quartz grains that make up the sandstone; ×150. (Courtesy of Oxford Cryosystems.)

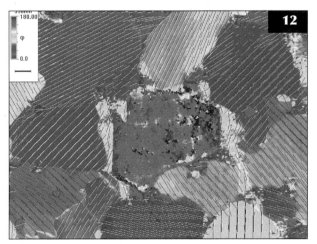

**12** Orientation ($\phi$) image of a sandstone sample obtained using the birefringence microscope imaging system. The lines show the orientation of the extinction (slow axis orientation) for each of the quartz grains; ×150. (Courtesy of Oxford Cryosystems.)

carried out on the sample. The various petrographic examination standards listed in *Table 2* give details of sampling requirements and provide guidance on appropriate sampling procedures (*Table 3*). For rock and building stone, the size of sample required depends on the grain size of the rock, with coarse-grained rock requiring larger samples than fine-grained rock. Rock/stone samples may consist of lump samples, drilled cores, or sawn blocks. For aggregate products it is important to ensure that a representative number of individual aggregate particles are examined. For this reason, the required sample mass increases as the nominal maximum particle size of the aggregate increases. When sampling concrete from a structure, broken fragments are not usually acceptable for petrography, as it is difficult to differentiate between inherent concrete defects and sampling damage. Diamond drilled concrete core samples (13) are preferred and the core diameter should be at least

**Table 3** Summary of sampling requirements for different geomaterials

| Material group | | Sample type/minimum size for petrographic examination |
|---|---|---|
| Building stone | | 100 mm × 100 mm × 50 mm sawn block or 5 kg lump sample |
| Roofing slate | | 1 whole slate |
| Rock | | 5 kg of lump or core samples |
| Coarse aggregate[1] | 63 mm (75 mm) nominal maximum size | 50 kg (180 kg) |
| | 31.5 mm (37.5 mm) nominal maximum size | 25 kg (90 kg) |
| | 16 mm (19 mm) nominal maximum size | 8 kg (45 kg) |
| | 8 mm nominal maximum size | 2 kg |
| Fine aggregate[1] | <4 mm (<4.75 mm) sized | 0.5 kg (23 kg) |
| Concrete | From structure | 100 mm diameter core |
| | From production run | 150 mm cube specimen |
| Concrete product | | 1 unit or 100 mm diameter core sample |
| Mortar, plaster, and render | From structure | 0.5 kg intact lump sample |
| Screed | From floor | 100 mm diameter core sample or 2 kg intact lump sample |
| Clay brick | From structure or production run | 1 unit |
| Ceramic tile | | 1 unit |
| Architectural glass | | 100 mm × 100 mm square piece |
| Bituminous mixtures | From pavement | 150 mm diameter core sample |
| | From production run | 20 kg |

[1]EN 932-3 requirements (with ASTM C295 requirements in brackets).

twice, and preferably three times, the maximum size of the coarse aggregate within the concrete. For construction products that are supplied as units, for example precast concrete or clay bricks, it is normally sufficient to examine one complete representative unit. Due consideration should be given to the health and safety issues that arise from sampling activities, including safe access to the sample locations.

Once obtained in the field, samples should be labelled indicating orientation details (bedding, outer surface, way up, and so on), placed into separate sealed sample bags, and each bag labelled with a unique identification reference. Concrete cores should first be wrapped in cling film to prevent carbonation. The exact sample locations should be recorded using a combination of written notes, drawings, and photography. These should include comprehensive inspection details of the location prior to sampling and an 'as-found' record photograph. These details should be made available to the petrographer if he/she was not present during the sampling operations.

## THIN SECTION SPECIMEN PREPARATION

High-power microscopical examination requires the preparation of petrographic 'thin section' specimens comprising ground slices of the sample mounted on glass slides, through which light will pass to allow microscopical observation. The thin section making technique was developed by geologists for the study of rocks. When applied to other construction materials that are soft, heat sensitive, and/or water sensitive, thin section preparation presents considerable challenges for the technician.

Thin section making starts with oven-drying the sample at a temperature that will not damage heat-sensitive materials (lower than 60°C). Once dry the specimen is vacuum impregnated with a low viscosity epoxy. The resin is usually coloured with a dye to assist in the determination of porosity and cracking when the thin section is microscopically examined. The colour of this dye would normally be fluorescent yellow except when investigating the thaumasite form of sulfate attack (TSA) in concrete, in which case blue dye is a better choice. Following impregnation the resin is heat cured in an oven or on a hot plate, again at <60°C. The cured sample block is then ground using a diamond cup wheel to expose the examination surface.

At this stage the 'thin-sectioning machine' is first used. The surface for examination is finely ground using the thin-sectioning machine, cleaned, and mounted on a glass slide using epoxy glue (either heat or ultraviolet light curing). The excess sample is then cut off using a special diamond saw with a vacuum chuck that holds the slide in place (14). All cutting and grinding must be done in oil or alcohol rather than water to avoid damage of water-sensitive materials. The thin section is then finely ground to the finished sample thickness on the thin-sectioning machine and finished by gluing a thin glass cover slip over the sample.

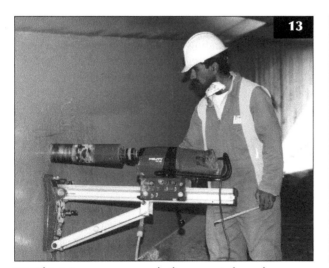

**13** Obtaining a core sample from a reinforced concrete structure by diamond drilling.

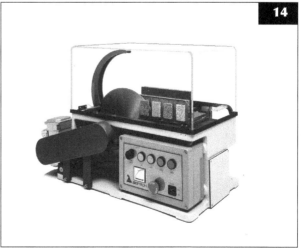

**14** A thin section cut-off and trim saw. (Courtesy of Logitech Materials Technologists and Engineers.)

There are two types of thin section making machines on the market. The first is the lapping machine (15), which grinds the sample to the required thickness on a lapping plate that is fed with abrasive grit (carborundum or aluminium oxide). The second type of machine grinds with fine diamonds embedded in a bronze wheel without the need for abrasive slurry. Lapping machines produce a finer finish (with 5 µm aluminium oxide power) than the diamond grinding wheel (typically utilizing 60 µm diamonds). Also, diamond grinding wheels will cause shattering of crystal grains within samples when operated incorrectly. On the other hand, diamond grinding wheel systems produce thin sections that are noticeably cleaner when viewed through the microscope.

The conventional thickness of thin section specimens is 30 µm, although this is occasionally varied depending on the application; for example, 20–25 µm thick specimens are sometimes used for concrete specimens as it makes it easier to resolve the details of the cement matrix. The plan size of thin section specimens tends to be either small (25 mm × 75 mm), medium (50 mm × 75 mm) or large (75 mm × 100 mm) (16). The size of slide required depends on the grain size of the material being thin sectioned, with small slides only being useful for fine-grained rocks and grouts. The majority of materials, including all concrete specimens, require the use of medium or large area slides.

It is essential that the petrographer has a thorough understanding of the thin section making process and the sample preparation materials. Ideally the petographer should have practical experience of thin-section preparation. Due care must be taken to ensure that artefacts of the thin section making process or storage conditions are not mistaken for natural features of the sample. Some of the more common pitfalls include:

- Over-heating during sample preparation may cause mineralogical changes, for example, carbonation of cements, or gypsum dehydrating to anhydrite.
- The use of water for cutting and grinding of water-sensitive samples may cause them to hydrate and/or dissolve away.
- Shrinkage microcracks caused by oven-drying prior to resin impregnation must not be confused with genuine microcracks in cementitious materials. Cement-rich concretes and mortars are especially prone to this.
- Open cracks and microcracks may become filled by glue (epoxy resin) used to stick down the cover slip. The glue is isotropic and could potentially be misidentified as isotropic alkali–silica gel in concrete, leading to an incorrect diagnosis of alkali–silica reaction (ASR). Glue filling microcracks can sometimes exhibit first-order interference colours and could potentially be misidentified as gypsum. In concrete this could lead to a misdiagnosis of sulfate attack (17).
- Cracks caused by shrinkage of the impregnating resin at the edge of specimens can easily be misidentified as surface delamination of the sample.
- Abrasives and grinding debris produced during the cutting and grinding processes can often be found filling cracks and voids in thin sections. Care must be taken to ensure that these are not mistaken for the products of deleterious reactions.

Suppliers of thin section preparation equipment and consumables are listed in Appendix A.

### FINELY GROUND SLICES AND HIGHLY POLISHED SPECIMENS

Finely ground slices are sometimes prepared to enhance the visual and low-power microscopical examination stage of petrographic examination. They are most commonly used in concrete petrography where a typical slice would be 100 mm square (minimum) by 25 mm thick. One surface of the slice would be finely ground on

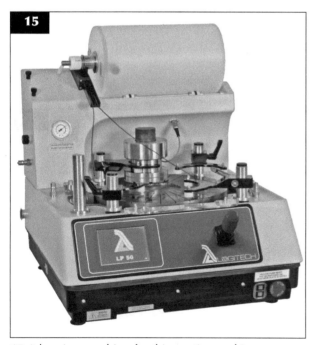

**15** A lapping machine for thin section making. (Courtesy of Logitech Materials Technologists and Engineers.)

a lapping machine (15) using carborundum or aluminium oxide abrasive power with a minimum particle size of 5 μm (F1000 grade).

Highly polished specimens differ from finely ground slices in that a finer surface finish is produced, and as this is difficult to achieve over a large area, the specimens are much smaller (typically <30 mm diameter round blocks). Samples for high-polishing are oven-dried and encapsulated in resin using a round mould. Once the resin has set the specimen is removed from the mould, roughly ground flat using a diamond cup wheel, and then polished using a polishing machine (18).

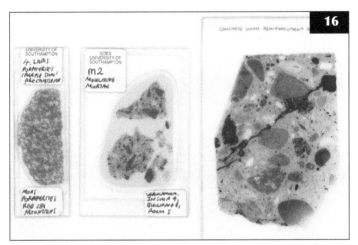

**16** Thin sections in three different sizes; small (left), medium (middle), and large (right).

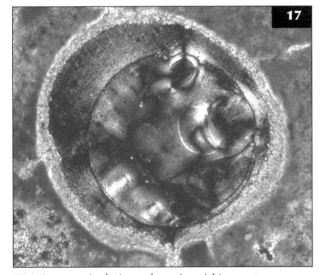

**17** Microscopical view of an air void in mortar, seen in thin section. The void is lined by resin used to consolidate the sample (dark green) and the void centre is filled by epoxy resin used to set the glass cover slip. The glue exhibits first-order interference colours; XPT, ×150.

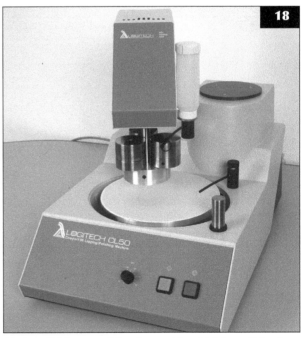

**18** A polishing machine for making highly polished specimens or polished thin sections. (Courtesy of Logitech Materials Technologists and Engineers.)

The polishing process utilizes successively finer grades of diamond polishing compounds routinely down to 1 µm, or even 0.25 µm for the finest work. Cementitious materials must be dried at less than 60°C and not exposed to water at any stage. Thin sections may be finished by high-polishing (called polished thin sections) and left without cover slip to allow observation in both transmitted and reflected light. Polished specimens are also required for supplementary investigation using SEM. Most highly polished specimens are usually water sensitive and it is good practice to store them in a desiccator, to avoid degradation from atmospheric moisture.

## STAINING AND ETCHING PROCEDURES

While most of the minerals present in geomaterials can be readily identified in thin sections, some are difficult to differentiate. These include, for example, the carbonate minerals of limestone, gypsum and anhydrite, feldspars, and clays in rocks. In addition, certain phases within cement clinkers and slags require etching. Techniques using chemical solutions or vapour selectively to etch or stain minerals as an aid to their identification are known to geological technicians and cement chemists. Details of staining and etching techniques are provided in the tables of Appendix B.

# Building stone

## INTRODUCTION

Stone, the primary geomaterial, has been used for building since man gave up the nomadic lifestyle of a hunter–gatherer and began to build permanent settlements. Natural stone is used in civil engineering for structural elements, facing and cladding, and hard landscaping. Current uses of natural stone in buildings include load-bearing and self-supporting masonry, masonry façades to framed buildings, cladding and lining, roofing, flooring, and miscellaneous elements.

The earliest stone structures were probably masonry constructed of unworked stone laid dry. Some ancient masons used stone blocks with close-fitting joints not requiring mortar. These often utilized blocks so large that they would not subsequently shift, only occasionally incorporating metal dowels for added stability. More commonly, less precisely worked masonry units were used, which had to be set in mortar to add stability to a wall or building and make it weatherproof. Traditional stone walls may be constructed from roughly squared blocks (random rubble) or with dressed regular layers (coursed). Building stone that has been selected, trimmed, and cut to specific shapes and sizes is referred to as 'dimension stone'. Traditional load-bearing masonry materials are inherently strong in compression and comparatively weak in tension. This has led to the development of masonry structures that generate only very small tensile stresses such as the arch, the vault, and the dome.

To take the required load, masonry walls for buildings more than three storeys high must be of great thickness and are consequently slow and expensive to construct. From the late 1800s these problems were overcome by the use of steel and reinforced concrete frames, which support the floors and walls on the beams of the frame. Modern cladding systems use relatively thin panels of stone that are individually attached to the frame with metal fixings.

Stone is also used as a prestige material for internal flooring and lining. Another significant application is stone roofing, with natural slate roofing being of particular importance. Slate investigations have become a substantial area of business for the petrographer in their own right and are therefore considered separately in Chapter 3. Road paving applications were historically one of the largest uses of stone in construction. Today the use of stone setts, kerbs, and paving is restricted to 'streetscape' projects where a particular appearance is required or for repairs of existing stone paving.

Of the immense variety of building stones that are available, all may be classified geologically as being in one of three broad groups. All stones are rock that are either of igneous, sedimentary, or metamorphic origin, depending on their composition and how they were formed. Each of these three groups will be discussed in detail in subsequent sections.

The marketing of stone has resulted in a wide variety of nonscientific stone names being introduced into the building stone industry. This can cause confusion when trying to determine the identity and source of particular stone types. The definitive name for any stone is its 'petrographic name', which is the geological identity determined by petrographic examination. Additionally, stones are likely to have a 'commercial name' for marketing purposes, which may relate to traditional stone names for the area of production or may be just a marketing tool. The stone industry groups stone types into 'commercial groups' as shown in *Table 4 (overleaf)*. In this chapter both the petrographic name and (if available) the commercial name will be included for each of the stone types discussed. The commercial name will be identified by single quotation marks.

The principal applications of petrographic examination to building stone investigations are:

- Initial suitability assessment of stone from new sources.
- Routine quality assurance of stone from the factory production run.
- Confirming the identity/quality of stone for purchasers.
- Diagnosing the causes of in-service deterioration/failure.
- Matching stone types for restoration of historic buildings.

# TESTING BUILDING STONE

Natural stone resources are variable in character and it is common to find the good stone of a certain locality passing within a short distance into less suitable material. Consequently, the fact that stone from a particular source has performed well in the past is not reliable proof that the currently produced stone is of the same quality. The quality of stone is also influenced by the methods of extraction, processing, and quality control that are used; these practices may change over the years. In addition, stone is now being imported from new sources across the world, primarily to exploit the cost savings of producing stone in developing counties with cheap labour. Therefore, there has never been a greater need for assurance of stone identity and for dependable prediction

of durability. Globally, stone testing requirements are dominated by two systems, the European standards (EN) and the American standards (ASTM).

In Europe, to ensure that stone is fit for purpose the Comité Européen de Normalisation (CEN) requires the manufacturer to undertake a programme of initial type testing to evaluate the composition, strength, and potential durability of every new source. Once it has been established that the stone is suitable for the proposed building stone application, further testing of the factory product is required at regular intervals for quality assurance. Further testing is sometimes required by the manufacturer, the purchaser, or third parties to investigate nonconformities/complaints, confirm identity, or to diagnose the causes of in-service failure. There are about 20 European standards covering test methods for

**Table 4** Stone classification: relationship between popular commercial and geological petrographic names (compiled from the terminology in EN 12670: 2002)

| Generic grouping | Petrographic name | | Commercial group |
|---|---|---|---|
| IGNEOUS | Basalt<br>Dolerite<br>Picrite | | BASALT |
| | Intrusive: | Granite<br>Diorite<br>Gabbro<br>Peridotite<br>Syenite | GRANITE |
| | Extrusive: | Rhyolite<br>Andesite | |
| METAMORPHIC | Gneiss | | |
| | Quartzite | | QUARTZITE |
| | Schist | | SCHIST |
| | Phyllite | | |
| | Slate | | |
| | | | SLATE |
| | Serpentinite<br>Marble | | MARBLE |
| SEDIMENTARY | Dolomite<br>Travertine<br>Limestone (polishable) | | |
| | Limestone (nonpolishable) | | LIMESTONE |
| | Sandstone | | SANDSTONE |

natural stone products. The standard suite of tests varies depending on the product type, in accordance with the product specifications listed below:

| Stone product specification | Applicable European standard |
|---|---|
| Masonry units | EN 771-6 |
| Exernal paving | EN 1341 |
| Setts | EN 1342 |
| Kerbs | EN 1343 |
| Rough blocks | EN 1467 |
| Rough slabs | EN 1468 |
| Cladding | EN 1469 |
| Modular tiles | EN 12057 |
| Flooring | EN 12058 |

All of the European product specifications require that petrographic examination be performed on stone from new sources as part of initial type testing and in many cases, at least every 10 years for existing sources, as part of factory production control. Stone products that have been subjected to standard suite of European tests can be CE marked. Currently, the European standards give little guidance as to how to interpret the test results and CE

marking does not necessarily mean that a stone will be fit for the intended purpose. In the absence of an approved classification system, the test data must be assessed by an engineer or stone specialist to determine if the stone is suitable for a particular project. Testing beyond that required for CE marking is prudent for certain applications.

In America, stone is specified in accordance with ASTM C1528 – *Selection of Dimension Stone for Exterior Use* (ASTM International, 2008a). The American system differs from the European one in that the stone test requirements are specific to the stone type rather than the product type. Each type of stone has its own ASTM standard as follows:

| Stone type | Applicable American standard |
|---|---|
| Granite | ASTM C615 |
| Limestone | ASTM C568 |
| Marble | ASTM C503 |
| Quartz-based | ASTM C616 |

The ASTM standards provide pass/fail criteria for a standard range of physical tests (*Table 5*). Petrographic examination is not currently a requirement of the ASTM standards, although it is likely to become one in the future.

**Table 5** Classification of stone quality from results of ASTM tests (modified from Burton, 1999)

| Stone type/ ASTM specification | Subdivision | Minimum density ASTM C97 (Mg/m$^3$) | Maximum absorption ASTM C97 (%) | Minimum compressive strength ASTM C170 (MPa) | Minimum flexural strength ASTM C880 (MPa) | Minimum modulus of rupture ASTM C99 (MPa) | Minimum abrasion resistance ASTM C241 |
|---|---|---|---|---|---|---|---|
| Granite/C615 | – | 2560 | 0.4 | 131 | 8.27 | 10.34 | 25 |
| Limestone/ C568 | Low density | 1760 | 12 | 12 | – | 2.9 | 10 |
| | Medium density | 2160 | 7.5 | 28 | – | 3.4 | 10 |
| | High density | 2560 | 3 | 55 | – | 6.9 | 10 |
| Marble/C503 | Calcite | 2595 | 0.2 | 52 | 7 | 7 | 10 |
| | Dolomite | 2800 | 0.2 | 52 | 7 | 7 | 10 |
| | Serpentinite | 2690 | 0.2 | 52 | 7 | 7 | 10 |
| | Travertine | 2305 | 0.2 | 52 | 7 | 7 | 10 |
| Quartz-based/ C616 | Sandstone | 2003 | 8 | 27.6 | – | 2.4 | 2 |
| | Quartzitic sandstone | 2400 | 3 | 68.9 | – | 6.9 | 8 |
| | Quartzite | 2560 | 1 | 137.9 | – | 13.9 | 8 |

## PETROGRAPHIC EXAMINATION AND COMPLEMENTARY TECHNIQUES

Within the European Union, petrographic examination of building stone is performed following a procedure given in EN 12407 (British Standards Institution, 2007). An American standard (WK2609) is currently undergoing development (ASTM International, 2006). Guidance may also be taken from the International Society for Rock Mechanics (1977) and BS 5930 (British Standards Institution, 1999). Petrographic examination comprises a combination of visual and low-power examination in hand specimen and more detailed high-power microscopical examination in thin section (and sometimes also polished specimen).

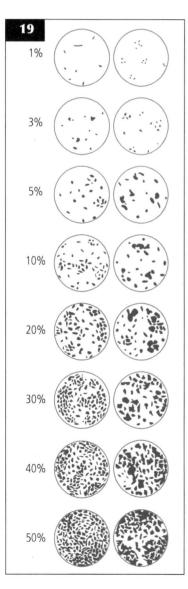

**19** Comparison chart for visual percentage estimation of constituents in thin section (after Terry & Chilingar, 1955).

Following arrival in the laboratory, the stone sample is first examined in the as-received condition using the unaided eye and low-power stereo-zoom microscope at magnifications of up to ×60. Stone features observed at this stage include colour, relative hardness, fabric, grain size, open and refilled cracks, pores, and cavities, and presence of macrofossils. The initial examination is used to determine the most appropriate location for thin sections (and possibly polished specimens) to be taken for further, more detailed high-power microscopical examination. Thin sections are prepared from slices cut across any bedding, rift or grain. Stone features determined in thin section include identification of constituent minerals and degree of weathering. If opaque minerals (such as pyrite, ilmenite, magnetite) are present, polished specimens may be prepared to enable their examination in reflected light. An atlas to aid the microscopical identification of opaque minerals is provided by Marshall *et al.* (2004).

The proportions of different minerals present within a stone may be approximately estimated by visual comparison with standard charts (**19**). The mineral proportions can be determined more accurately by point counting. If the microscopical examination proves insufficient to assign a petrographic definition of the stone, further investigation may be necessary. Suitable techniques include mineralogical analysis by XRD, which provides semiquantitative estimates of mineral proportions, and XRF analysis, which gives accurate elemental analyses.

## STONE FROM IGNEOUS ROCKS

Igneous rocks originate as molten magma far down in the earth's mantle, rising towards the surface before they solidify. The rock masses which are produced by igneous activity can be classified in three groups, according to their mode of emplacement:
- *Intrusive rocks:* dykes, sills, plugs, plutons, and batholiths formed by consolidation of magma in spaces within the crust.
- *Extrusive rocks:* lava flows formed by magma erupted from volcanoes at the earth's surface.
- *Pyroclastic rocks:* agglomerate, tuff, and ignimbrite formed by the accumulation at the earth's surface of dust and fragments ejected by explosive volcanic eruptions.

The minerals which crystallize from a cooling magma depends on which elements are present in the magma. The mineralogical composition of any igneous rock is therefore closely related to its chemical composition, and both characteristics are used to classify igneous rocks. In terms of chemical composition, the principal component of all igneous rocks is silica ($SiO_2$). Igneous rocks are usually grouped according to their silica content as follows:

|  | $SiO_2$ (%) |
|---|---|
| Acid | >66 |
| Intermediate | 52–66 |
| Basic | 45–51 |
| Ultrabasic | <45 |

Acid igneous rocks have a high content of quartz and feldspars that are rich in silica, alkalis (sodium and potassium), and aluminium. Basic rocks are rich in minerals that contain iron, magnesium, and calcium (collectively called ferromagnesian minerals). To the four chemical groups (acid, intermediate, basic, and ultrabasic) must be added a fifth: that of the alkaline rocks, in which the silica content varies from basic to acid and the content of alkalis is higher than the other four groups. The naming of igneous rocks also depends on rock texture. The main divisions are drawn between the lavas and associated rocks, which have fine-grained or glassy textures, the medium-grained rocks formed in small intrusive bodies, and the coarse-grained rocks of large or slowly cooled intrusions. A classification of igneous rock for engineering purposes is provided in *Table 6*.

**Table 6** Geological classification of igneous rock for engineering purposes (a modified form of this table appears in BS 5930:1999)

| Rocks with massive structure and crystalline texture | | | | | Bedded rocks (with at least 50% of grains of volcanic origin) | Grain size (mm) |
|---|---|---|---|---|---|---|
| Grain size description | Pegmatite | | | | Fragments of volcanic ejecta in a finer matrix | >20 |
| Coarse | **Granite** | **Diorite Syenite** (alkaline) | **Gabbro** | **Pyroxenite Peridotite** | **Agglomerate** Rounded grains **Volcanic breccia** Angular grains | 2–20 |
| Medium | **Micro-granite** | **Micro-diorite** | **Dolerite** (or diabase) | | Cemented volcanic ash **Tuff** | 0.06–2 |
| Fine | **Rhyolite** | **Andesite Dacite Trachyte** (alkaline) | **Basalt** | | **Fine-grained tuff** | 0.002–0.06 |
| | | | | | **Very fine-grained tuff** | <0.002 |
| Glassy | **Obsidian** | **Volcanic glass** | | | | Amorphous or crypto-crystalline |
| | Pale ← colour → Dark | | | | | |
| | Acid (much quartz) | Intermediate (some quartz) and Alkaline (much sodium and potassium) | Basic (little or no quartz) | Ultrabasic | | |

## ACID IGNEOUS ROCK

Acid igneous rocks occur mainly as the coarsely crystalline plutonic variety known as granite and to a much lesser extent, finely crystalline rhyolite lavas. Rhyolite lavas are often associated with pyroclastic rock such as tuff and ignimbrite, which may also have an acid composition.

Granite is a very important building stone that is popular for its pleasant appearance, strength, and durability. Granite is composed essentially of feldspar (40–80%, mainly alkali feldspar with some plagioclase), quartz (20–60%), and dark-coloured ferromagnesian minerals (>20%, typically biotite mica or hornblende). Accessory minerals (which are present in such small quantities that they are not considered for rock classification purposes) may include augite, muscovite mica, apatite, zircon, and magnetite.

Unweathered granite is available in a range of light colours including white, grey, pink, and red. The dominant colour of fresh granite is nearly always determined by that of the feldspars, which are the most abundant constituent. Figure 20 shows a red hornblende granite that has been quarried at Aswan, Egypt since 4000BC and was used extensively in Pharaonic architecture (Arnold, 1991). Figure 21 shows a white granite that is currently being produced in China. Discolouration caused by weathering results in brownish and yellowish coloured granites, which are also produced for sale. It is important to establish the weathering grade of the granite as this could potentially affect its strength and durability. Figure 22 shows a yellow coloured granite that has been slightly discoloured and slightly weakened by weathering (grade II). Rock weathering is discussed in Petrography of stone defects and

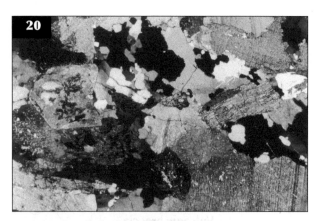

**20** Aswan granite (Egypt) consisting chiefly of plagioclase feldspar (lower right), orthoclase feldspar, and quartz (upper centre), with minor hornblende (brown, centre left) and biotite mica (upper right). From the casing of Menkaure's pyramid at Giza; XPT, ×35.

**21** 'Royal White' granite (China) consisting chiefly of alkali feldspar (lower left), plagioclase (upper left), and quartz (upper right), with minor biotite mica (green/brown, centre); XPT, ×35.

**22** 'Yellow Rock' granite (China) consisting chiefly of alkali feldspars that exhibit perthitic intergrowth (centre) and quartz (lower right), with minor biotite mica (pink, lower centre). This stone has been slightly weathered giving it a brown discolouration from iron compounds (best viewed in PPT); XPT, ×35.

decay, p. 47, and Soundness, impurities, and undesirable constituents of aggregates, p. 66, and a weathering classification scheme is shown in *Table 14*, p. 67.

Most granites used for building are coarsely crystalline and even textured. Some exhibit porphyritic texture with large feldspar crystals set in a finer groundmass. An extreme and attractive example of this is Rapakivi texture, where large oval orthoclase feldspars occur with a mantle of sodium-rich plagioclase feldspar (23).

Rhyolite lavas are little used as building stone. They consist of phenocrysts of quartz and alkali feldspar in a finely crystalline or glassy groundmass. When acid lavas cool very rapidly they do not crystallize and instead, solidify to form a natural glass, obsidian (24). This was historically worked for use as a gemstone and for tool making, and is important archaeologically.

Tuffs of acid and intermediate composition are used for building, especially for hard landscaping. Figure 25 shows a tuff of rhyolitic composition that is marketed as 'porphyry' for use as paving.

### INTERMEDIATE IGNEOUS ROCKS

Although they are less used for building than acid or basic rocks, intermediate igneous rocks include some significant and attractive stones. Syenite is a coarsely crystalline plutonic intermediate rock consisting chiefly of alkali feldspar with less than 5% quartz and/or feldspathoid. Clinopyroxene, hornblende, biotite mica, or olivine may be present in minor proportions. Larvikite, a syenite from Norway, is a popular decorative stone because of the moonstone schiller shown by its alkali feldspars.

**23** 'Baltic Brown' Rapakivi textured granite (Finland) showing a section through the edge of a large feldspar orb consisting of orthoclase feldspar (right) mantled with sodic plagioclase feldspar (centre). The coarsely crystalline groundmass is seen (left) including quartz (grey/white), biotite mica (brown/green), and hornblende (brown); XPT, ×35.

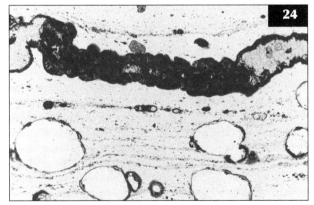

**24** Obsidian (Lipari, Italy) consisting of isotropic glass that exhibits flow banding, air bubbles, and spherulites (brown); PPT, ×35.

**25** 'Porfido Rosso' (Italy). A rhyolitic tuff comprising phenocrysts of quartz and feldspar set in a groundmass of cryptocrystalline devitrified volcanic glass (black); PPT, ×35.

Figure 26 shows a popular, currently available larvikite stone. Figure 27 shows a syenite that was used in antiquity for building the remarkable monuments of the Aksumite civilization in Ethiopia.

Another igneous rock of intermediate composition is the 'Roman Imperial Porphyry', a dark purple stone that was much sought after in the Roman and Byzantine worlds (Maxfield & Peacock, 2001). Imperial Porphyry is a dacite found only at one location in the Red Sea Mountains of Egypt. The rock is porphyritic with phenocrysts of plagioclase feldspar (oligoclase to andesine) and augite, set in a very fine groundmass of feldspar (Klemm & Klemm, 2008). The most famous variety 'Lapis Porphyrites' (28) has white feldspars set in a deep purple groundmass (coloured by the presence of haematite). 'Rubet Porphyrites' has pink feldspars (coloured by the presence of the pink mineral piemontite) set in a deep purple groundmass (29). Another variety,

**26** 'Blue Pearl' (Norway) syenite stone consisting chiefly of cryptoperthitic alkali feldspar (grey) with minor clinopyroxene (lower left, brightly coloured), biotite mica (brown/green), and opaque minerals (centre, black); XPT; ×35.

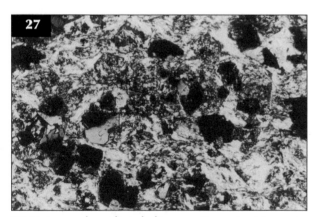

**27** Metamorphosed nepheline syenite comprising phenocrysts of pyroxene (brightly coloured) and amphibole (brown) in a finely crystalline groundmass of nepheline and alkali feldspar. Ghosts of nepheline and feldspar phenocrysts can still be seen (from The Great Stela, Aksum, Ethiopia); XPT, ×35.

**28** 'Lapis Porphyrites' from Mons Porphyrites, Egypt. Porphyritic dacite comprising phenocrysts of feldspar (white) and augite (brown), in a groundmass of very finely crystalline feldspar (grey); XPT, ×35.

'Lapis Porphyrites Niger' (30) has white feldspars in a groundmass that appears black with a greenish tinge.

## BASIC IGNEOUS ROCKS

The basic rocks most used for building include the coarsely crystalline plutonic variety known as gabbro and the medium crystalline variety called dolerite (known as diabase in America) that occur in dykes and sills. These rocks have a dark appearance ranging from dark grey to black and are popular decorative and cladding stones. They are often incorrectly named in the stone industry as 'black granite'. In fact they contain much less silica, being composed essentially of feldspar (mainly plagioclase feldspar with minor alkali feldspar) and dark coloured ferromagnesian minerals (>20%, typically pyroxene, olivine, biotite mica, or hornblende), and only accessory amounts of quartz. Figure 31 shows a stone of typical hypersthene gabbro composition while Figure 32 shows a

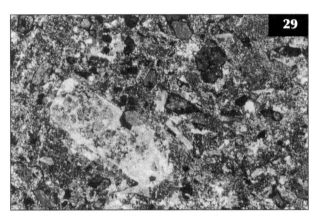

**29** 'Rubet Porphyrites' from Mons Porphyrites, Egypt. Porphyritic dacite comprising phenocrysts of feldspar (white) and augite (brown), in a groundmass of very finely crystalline feldspar (grey). The feldspar phenocrysts have inclusions of piemontite (pink); XPT, ×35.

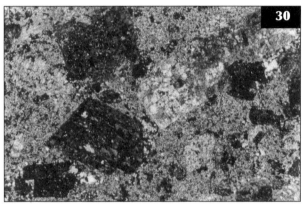

**30** 'Lapis Porphyrites Niger' from Mons Porphyrites, Egypt. Porphyritic dacite comprising phenocrysts of feldspar (grey/white) in a groundmass of very finely crystalline feldspar (grey) and hornblende (green/brown); XPT, ×35.

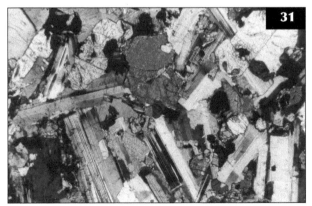

**31** 'Impala Black' gabbro (South Africa) consisting chiefly of coarsely crystalline plagioclase feldspar (grey), orthopyroxene (brown, upper right), and clinopyroxene (brightly coloured), with minor biotite mica (green) and magnetite (black); XPT, ×35.

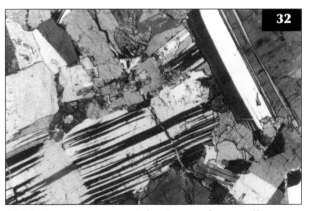

**32** Norite consisting mainly of coarsely crystalline calcic plagioclase (labradorite composition) feldspar (grey) and orthopyroxene (brown), with minor clinopyroxene (blue). A natural joint that is tightly filled with silica runs obliquely through the field of view; XPT, ×35.

norite, which is a gabbro that includes orthopyroxene but very little clinopyroxene. Figures **33** and **34** both show dolerite stones, with the first having a porphyritic texture and the second having the more typical uniform medium crystalline texture. Figure **35** shows a dolerite that is green (rather than black) in hand specimen, caused by the presence of the green alteration mineral, chlorite.

The finely crystalline equivalent of gabbro is basalt, which occurs as lava. It has the same black appearance and is used locally for masonry and hard landscaping. In terms of texture, basalt may exhibit relatively uniform crystal size, or may be porphyritic (**36**). A lava flow that contains considerable trapped gas vesicles may develop scoriaceous texture (**37**).

## ULTRABASIC IGNEOUS ROCKS

Ultrabasic rocks consist mainly of ferromagnesian minerals, such as pyroxene and olivine, and are dark

**33** 'Crystal Black' porphyritic dolerite (China) chiefly comprising phenocrysts of augite (pink/yellow/blue) set in a groundmass of plagioclase (labradorite composition) feldspar (grey/black), with minor olivine (green/brown); XPT, ×35.

**34** 'Zimbabwe Black' dolerite (Zimbabwe) consisting chiefly of medium-crystalline plagioclase feldspar (grey), orthopyroxene (grey/brown), and clinopyroxene (brightly coloured), with minor biotite mica (dark brown) and magnetite (black); XPT, ×35.

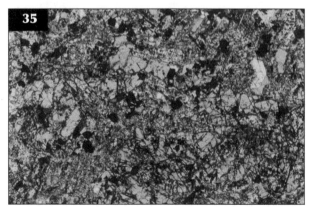

**35** 'Royal Green' dolerite (China) that exhibits alteration of ferromagnesian minerals to chlorite (green); PPT, ×35.

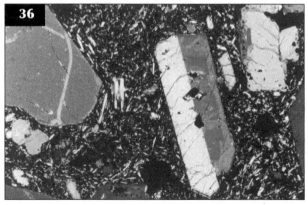

**36** Porphyritic basalt (Equatorial Guinea) comprising phenocrysts of augite (green/orange, centre) and olivine (pink) set in a finely crystalline groundmass of plagioclase feldspar (black/grey); XPT, ×35.

coloured. Fresh ultrabasic rocks such as peridotite are little used in building. However, certain types of altered ultrabasic rocks are used as decorative stone. The alteration process known as serpentinization alters ferromagnesian minerals to the mineral serpentine, and extensively serpentinized rocks are termed serpentinite (38, 39). These best known varieties are green coloured (often called 'Verde') and may have a mottled appearance. Serpentinites take a polish well but are relatively soft, so they are most suitable for decorative panels or carving. Altered ultrabasic rocks may contain asbestos, which makes them a potential health hazard (40). The asbestos is usually well bound into the stone and usually represents no danger in-service. However, the possibility of asbestos fibres being released should be considered for stone processing and building demolition.

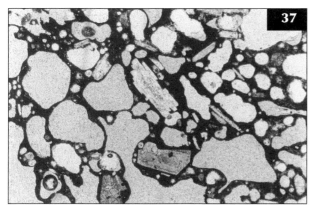

37 Basaltic scoria showing characteristic vesicular texture (from Mount Etna, Sicily); PPT, ×35.

38 Serpentinite consisting wholly of the mineral serpentine (from Cyprus); XPT, ×25.

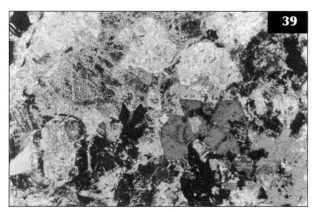

39 'Polyphant Stone' from Cornwall, England. A serpentinized picrite consisting of talc (pink), serpentine (dark blue/grey/green), and carbonate minerals (light brown); XPT, ×35.

40 Serpentinite (Cyprus) exhibiting a vein of chrysotile (white asbestos, fibrous, brightly coloured). Serpentine appears grey; XPT, ×150.

## STONE FROM SEDIMENTARY ROCKS

Sedimentary rocks are formed from fragments of rock eroded from the weathered land surface, transported by water, wind, and ice, and deposited in coastal plains, deltas, and seas. The loose sediments are deposited in layers (beds) which become compacted and cemented to form bedded rock. Detrital sediments are formed by the accumulation of transported particles of old minerals, and of new minerals, produced during weathering. These include sandstones, mudstones, and conglomerate. Chemical–organic sediments are formed from materials transported in solution and from materials manufactured by plants and animals, such as limestone, dolomite, ironstone, and evaporites. The two main characteristics used to classify sedimentary rocks are the grain size and composition. Sedimentary rocks are widely used for building, with sandstones and limestones being the

**Table 7** Geological classification of sedimentary rocks for engineering purposes (a modified form of this table appears in BS 5930:1999)

| Grain size (mm) | Bedded rocks | | | | |
|---|---|---|---|---|---|
| >20 | Grain size description | | | At least 50% of grains are of carbonate | Chemical and carbonaceous rocks |
| 2–20 | Rudaceous | | **Conglomerate** Rounded boulders, cobbles, and gravel cemented in a finer matrix **Breccia** Irregular rock fragments in a finer matrix | **Limestone** **Magnesian limestone** **Dolomite** | **Halite** **Anhydrite** **Gypsum** |
| 0.6–2 | Arenaceous | Coarse | **Sandstone** Angular or rounded grains, commonly cemented by clay, calcitic, or iron minerals | | |
| 0.2–0.6 | | Medium | **Orthoquartzite** Quartz grains and siliceous cement **Arkose** Many feldspar grains | | |
| 0.06–0.2 | | Fine | **Greywacke** Many rock chips | | |
| 0.002–0.06 | Argillaceous | | **Mudstone** / **Siltstone** Mostly silt | **Chalk** | |
| <0.002 | | | **Shale** Fissile / **Claystone** Mostly clay | | |
| Amorphous or cryptocrystalline | | | **Chert** Occurs as nodules and beds in limestone and calcareous sandstone **Flint** Occurs as bands of nodules in the Cretaceous Chalk of western Europe | | **Coal** **Lignite** |

predominant types used. A classification of sedimentary rocks for engineering purposes is provided in *Table 7*.

## SANDSTONE

Sandstones are formed of lithified quartz grains between 0.06 mm and 2 mm in size. A small proportion of other mineral grains such as feldspar, mica, and rock fragments may also be present. The grains may be closely packed together forming a compact stone or sparsely distributed to form a porous stone. The grains are cemented together either by silica, calcite, iron compounds, or clay minerals. The strength and durability of sandstone varies considerably depending on the size of the grains, grain packing, porosity, and nature of the cement. Sandstones are available in a range of colours including white, grey, brown, orange, pink, and red, which is determined by their mineralogy and degree of weathering.

Figure 41 shows a strong and durable grey/brown quartzitic sandstone ('York Stone') that is used for paving. It consists of interlocking quartz grains, cemented by silica and has relatively low porosity (6%). Figure 42 shows a softer 'Nubian Sandstone' (Cretaceous) that was used for the construction of ancient Egyptian temples. It consists of well sorted, fine-sized quartz grains giving a relatively high intergrain porosity (20%). Figure 43 shows a sandstone that has a dark grey to blue/green colour from the inclusion of carbonaceous matter and chlorite in the matrix between quartz grains. Figure 44 shows a calcareous sandstone with calcite cement. This sandstone is pink in hand specimen due the presence of red iron

**41** 'Northowram' quartzitic sandstone (Halifax, England) consisting mainly of quartz grains (grey/white/black) with traces of muscovite mica (elongated, brightly coloured) and clay minerals (brown); XPT, ×150.

**42** Nubian sandstone (Egypt) consisting of quartz grains (white) and clay minerals (brown). Macropores are shown (yellow); PPT, ×35.

**43** 'Blue Pennant' sandstone (Wales) consisting of quartz grains (grey/black) with a matrix of clay minerals (bright) and carbonaceous matter (brown); XPT, ×150.

**44** 'Cambusmore' sandstone (Scotland) consisting of quartz grains (grey/black/white) with a veneer of haematite (red). The stone is cemented by calcite (pink); XPT, ×150.

compounds around the quartz grains. Figure **45** shows 'Caithness Stone' which is a durable paving stone of dark grey colour. It comprises bands of fine sandstone and mudstone with matrix/cement of calcite and phyllosilicate minerals.

## LIMESTONE

Limestones are made largely of calcium carbonate ($CaCO_3$), usually in the form of calcite. They are of many kinds and they exhibit a wide range of strength and durability. Organic limestones are formed from fragments of calcareous animal remains such as shells and coral. Chemical limestones are formed by precipitation of $CaCO_3$ in seawater or at hot springs on land. Clastic limestones are formed from eroded and redeposited fragments of pre-existing limestones. Dolomitic limestones contain a >10% proportion of the mineral dolomite, $CaMg(CO_3)_2$.

There are two widely used classification systems for limestones. In Dunham's classification (1962) rocks are assigned names according to their depositional texture, which is primarily related to the energy of the depositional environment *(Table 8)*. The Folk (1959, 1962) limestone classification assigns names primarily according to the proportions of different allochems and the nature of the cement (*Table 9*). Allochems are grains or particles found in most limestones such as bioclasts, ooliths, peloids, or intraclasts. The spaces between allochems may be wholly or partially filled by a matrix/cement of micrite (carbonate mud, <4 µm grain size), microsparite (fine calcite crystals, 4–30 µm size) or sparite (coarse calcite crystals, >30 µm size). The presence of macropores (>5 µm size) and micropores (<5 µm size) is an important factor which greatly influences durability of limestone used for building. Macropores typically occur as interparticle pores between allochems, while micropores typically occur as intraparticle

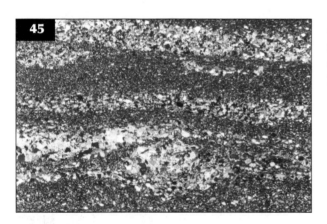

**45** 'Caithness Stone' (Scotland) comprising layers of fine sandstone with quartz grains (grey) and calcite cement (pink) interbedded with mudstone (brown); XPT, ×35.

**Table 8** Dunham (1962) scheme for classification of limestone

| Original components not organically bound together during deposition | | | | Components organically bound during deposition |
|---|---|---|---|---|
| Contains carbonate mud | | | No carbonate mud | |
| Mud-supported | | Grain-supported | | |
| <10% allochems | >10% allochems | | | |
| **Mudstone** | **Wackestone** | **Packstone** | **Grainstone** | **Boundstone** |

pores within the allochem fabric and intercrystal pores within the cement and allochems.

Limestone is used for building locally and certain varieties are exported around the world. It has provided the fabric for some of the most magnificent buildings in Britain. The more important British limestones will be discussed below with a selection of limestones from other countries following.

In northern England, the carboniferous limestone provides many hard, compact building stones that are usually cut and polished and sold as panels or tiles. Figure 46 shows 'Swaledale Fossil' limestone that is a pale brown with many attractive fossils (packed biomicrite). It is used for mainly decorative flooring and cladding.

The Jurassic System has provided more building limestone than any other in Britain. Most of the Jurassic limestones used for building are dominantly oolitic with variable proportions of shell debris. The Lincolnshire limestone (Bajocian, Middle Jurassic) has been the source

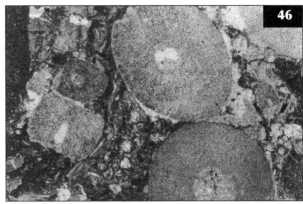

**46** 'Swaledale Fossil' limestone (Yorkshire, England) comprising grain-supported fossil allochems (mainly crinoids) composed of sparry calcite, with a matrix of micrite; XPT, ×35.

**Table 9** Folk (1959, 1962) scheme for classification of limestone

| Volumetric allochem composition | | | >10% allochems | | <10% allochems | | | |
|---|---|---|---|---|---|---|---|---|
| | | | Sparry calcite> Micrite | Micrite> Sparry calcite | 1–10% allochems | <1% allochems | | Undisturbed reef and bioherm rocks |
| >25% intraclasts | | | **Intrasparite** | **Intramicrite** | Intraclasts **Intraclast-bearing micrite** | | | |
| <25% intraclasts | <25% ooids | >25% ooids | **Oosparite** | **Oomicrite** | Ooids **Ooid-bearing micrite** | Micrite, or if sparry patches present **dismicrite** | | |
| | | Volume ratio, bioclasts:peloids >3:1 | **Biosparite** | **Biomicrite** | Bioclasts **Fossiliferous micrite** | | | |
| | | 3:1 to 1:3 | **Biopelsparite** | **Biopelmicrite** | | | | |
| | | <1:3 | **Pelparite** | **Pelmicrite** | Peloids **Peloid-bearing micrite** | | | **Biolithite** |

Most abundant allochems

of several of England's most famous building stones, including 'Ancaster', 'Barnack', 'Clipsham', and 'Ketton'. They range in colour from pale whitish-brown to yellowish-brown. Figure **47** is a general view of 'Ketton' limestone, which is a classic oolitic limestone, being highly porous and consisting entirely of ooliths. Ooliths in limestones are rounded grains that form by accretion of calcium carbonate around a nucleus. Figure **48** is a close view of 'Ketton Stone' showing the concentric ring structure of the ooliths. Figure **49** shows 'Ancaster Stone', a porous, cream–buff-coloured oolitic limestone (oobiosparite). Figure **50** shows 'Stamford Stone', a pale buff-coloured, highly porous, bioclasic oolite limestone. Figure **51** shows the blue/cream variety of 'Clipsham Stone', which consists mainly of intraclasts that are well cemented by sparite.

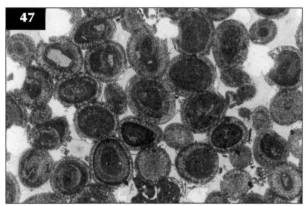

**47** 'Ketton' limestone (Lincolnshire, England) consisting entirely of grain-supported ooliths (brown) and with a high proportion of macropores (20%, yellow/white); PPT, ×35.

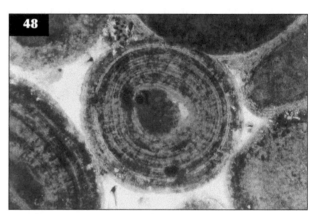

**48** Close view of 'Ketton' limestone showing concentric ring structure of a calcareous oolith. The brighter rings are more microporous than the darker ones; UV, ×150.

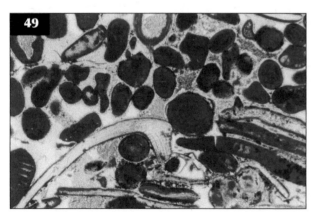

**49** 'Ancaster' limestone (Lincolnshire, England) consisting of grain-supported ooliths (dark brown) and minor shell fragments, with a sparry calcite matrix (pink) and a significant proportion of macropores (15%); XPT, ×35.

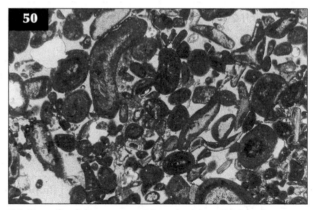

**50** 'Stamford' limestone (Lincolnshire, England) consisting of grain-supported shell fragments and ooliths (brown), and with a high proportion of macropores (20%, yellow); PPT, ×35.

'Bath Stone' (Bathonian, Middle Jurassic) is a generic term for the pale to dark yellowish-brown, oolitic limestones quarried and mined around the city of Bath (Somerset, England). Figure 52 shows 'Monks Park' limestone (oosparite), a variety of 'Bath Stone' that consists mainly of ooliths that are well cemented by sparite.

'Portland Stone' (Portlandian, Upper Jurassic) is perhaps the best known and possibly the most widely used of England's building stones. It is an even-grained, buff-coloured, porous oolitic limestone (oolitic grainstone) that weathers to a white colour on buildings. Three main beds are still quarried, the 'Whitbed' (53, 54), 'Basebed', and the

**51** 'Clipsham Blue/Cream' limestone (Lincolnshire, England) consisting of intraclasts (red, nonferroan microcrystalline calcite) cemented by ferroan sparry calcite (blue) with no macropores. The thin section has been stained in accordance with Dickson's method; PPT, ×150.

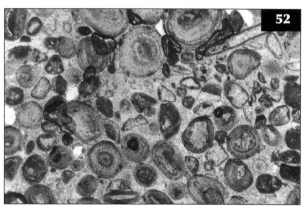

**52** 'Monks Park' limestone (Wiltshire, England) consisting of ooliths (pink, nonferroan microcrystalline calcite) cemented by ferroan sparry calcite (blue) with no macropores. The thin section has been stained in accordance with Dickson's method; PPT, ×35.

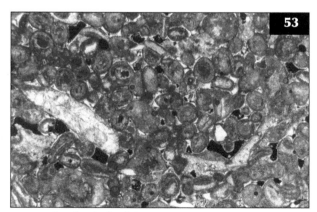

**53** 'Portland Whitbed' limestone (Dorset, England) consisting chiefly of grain-supported ooliths (brown) with traces of shell, and with a high proportion of macropores (10%, yellow); XPT, ×35.

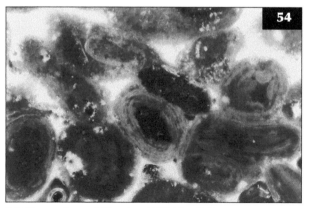

**54** Close view of 'Portland Whitbed' limestone showing differences in porosity. The brightest areas are more porous; UV, ×150.

'Roach'. The latter is notable for the presence of cavities formed from distinctive gastropod fossils (55).

'Purbeck Stone' (Upper Jurassic) is a hard shelly limestone that takes a good polish and is mainly used for decorative applications and paving. Figure 56 shows a coarse-grained variety known as 'Purbeck Grub' and Figure 57 shows a finer-grained variety called 'Purbeck Cap'.

The Cretaceous System has yielded a number of limestones that are important locally in southern England. The Chalk (Upper Cretaceous Chalk) is a very fine-grained, very pure, microporous limestone ranging from white to grey in colour (58). Most chalk is soft and weathers easily, but is sometimes used for building interiors in southern England. Certain harder beds within

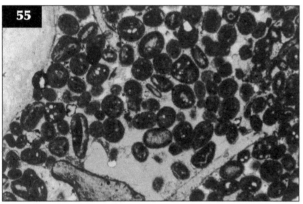

55 'Perryfield Roach' limestone (Dorset, England) consisting chiefly of grain-supported ooliths (brown) with minor shell, and with a high proportion of macropores (10%, yellow). This view includes one of the cavities that are characteristic of 'Roach', which originated as empty moulds of fossils; PPT, ×35.

56 'Purbeck Grub' biosparite limestone (Dorset, England) consisting of coarsely crystalline calcite (sparite) and shell bioclasts, with a sparite cement; XPT, ×35.

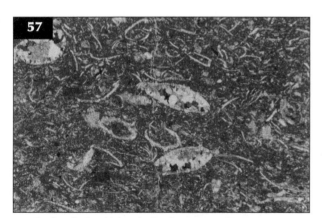

57 'Purbeck Cap' biomicrite limestone (Dorset, England) consisting of sparry calcite shell bioclasts (pink) with a matrix of microcrystalline calcite (micrite, brown); XPT, ×35.

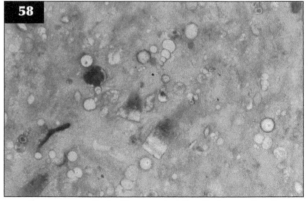

58 Close view of Chalk (Cambridgeshire, England) showing its high microporosity. Sparse biomicrite consisting of microcrystalline calcite (micrite) with minor proportions of microfossils (mainly foraminifera); UV, ×150.

the chalk, referred to as 'clunch' are more suitable for building (Clarke, 2004). Figure **59** shows 'Totternhoe Stone', a clunch variety that is durable enough for use in external walling. Figure **60** shows the slightly finer grained 'Cambridgeshire Clunch', which is more widely used for intricately carved, internal decorative stone work and vaulting.

Many popular building limestones are produced in Europe. Figure **61** shows 'Anstrude Stone' (Bathonian, Jurassic), a buff-coloured oolitic limestone (oobiosparite) from France, which also contains crinoid debris and has calcite cement. Figuire **62** shows 'Buxy Stone' (Bajocian, Jurassic), a beige/yellowish–red/grey compact crinoidal

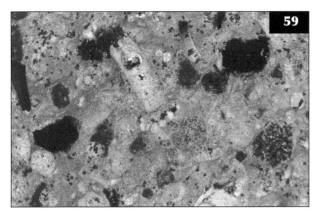

**59** 'Totternhoe Clunch' (Bedfordshire, England) limestone (packed biomicrite) consisting of shell fragments (pink) cemented by microcrystalline calcite (brown) with minor proportions of collophane fragments (black) and traces of glauconite (green); XPT, ×150.

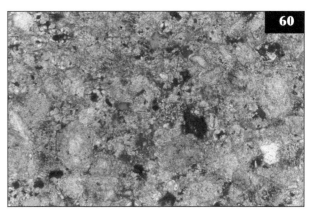

**60** 'Cambridgeshire Clunch' (Cambridgeshire, England) limestone (packed biomicrite) consisting of shell fragments (pink) cemented by microcrystalline calcite (brown) with traces of quartz grains (grey), collophane fragments (black), and glauconite (green); XPT, ×300.

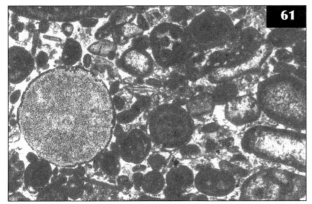

**61** 'Anstrude' limestone (France) consisting of ooliths and bioclasts (including crinoids, orange/left) with a cement of sparry calcite (white); PPT, ×25.

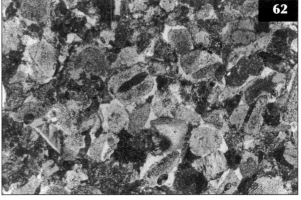

**62** 'Buxy Bayadere' limestone (France) consisting of bioclasts (mainly crinoids) that are well cemented by sparry calcite; XPT, ×35.

limestone (biosparite). Figure **63** shows 'Petit Granit' (Carboniferous), a hard polishable, grey fossiliferous limestone from Belgium. Figure **64** shows 'Jura' limestone, which is a hard compact, beige biopelmicrite. Figure **65** shows 'Caliza Capri' from Spain, which is a cream/buff limestone (oobiosparite). Figure **66** shows 'Moleanos' from Portugal, which is a compact, buff-coloured limestone (oosparite). Figure **67** shows 'Rossa Verona' from Italy, which is a hard polishable, orange limestone (biosparite) with occasional stylolites.

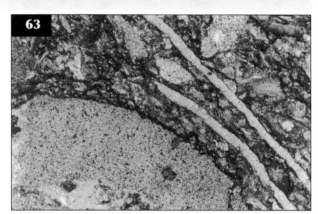

**63** 'Petit Granit' limestone (Belgium) consisting of bioclasts (mainly crinoids and shells) that are well cemented by sparry calcite; XPT, ×35.

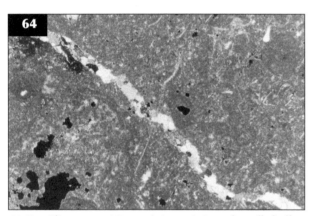

**64** 'Jura' limestone (Germany) consisting of small shell fragments (pink) and micritic peloids (brown), cemented by microcrystalline calcite (brown). A vein of coarsely crystalline calcite is seen in this view; XPT, ×35.

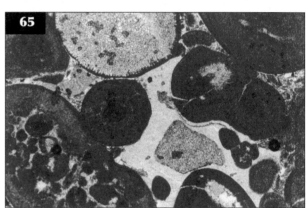

**65** 'Caliza Capri' limestone (Spain) consisting of rounded ooliths and bioclast allochems (brown) cemented by sparry calcite (pink); XPT, ×35.

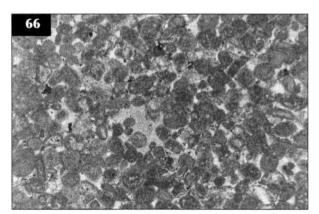

**66** 'Moleanos' limestone (Portugal) consisting of ooliths (brown) cemented by sparry calcite (pink); XPT, ×35.

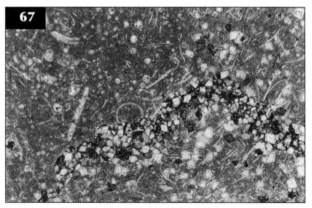

**67** 'Rossa Verona' limestone (Italy) consisting of fine shell fragments supported by a matrix of microcrystalline calcite (brown). A stylolite running across the field of view is filled by red iron compounds and rhomb-shaped dolomite crystals; XPT, ×35.

No section on building limestones would be complete without mentioning the Pyramids of Giza in Egypt. The cores of the pyramids were built from blocks of the local Eocene limestone (Mokattam Formation). Figure **68** shows this fine-grained, buff/cream-coloured biomicrite that comprises 30–50% foraminifera microfossils (mainly Nummulities) and 50–70% micrite matrix. The casing stones for the pyramids were quarried at Tura on the opposite (east) side of the River Nile. Here, the same rock (Mokattam Formation) is finer-grained and it was for this quality (and its ability to form tighter joints) that it was selected for use as casing blocks. Figure **69** shows this 'Tura' limestone, which is a very fine-grained, sparse biomicrite comprising approximately 10% microfossil fragments and 90% micrite matrix. In 1974 a radical theory was proposed that involved the pyramid casing stones being cast *in situ* as geopolymer 'concrete', rather than being natural stone from the Tura quarries. This theory has recently been disproved by a petrographic study (Jana, 2007).

Dolomite (also called magnesian limestone) is named after the mineral dolomite $CaMg(CO_3)_2$ that it contains. It is formed from limestone by chemical reactions during diagenesis. Dolomitic rocks are classified according to their percentage content of the dolomite mineral as follows:

- 0–10% dolomite = limestone.
- 10–50% dolomite = dolomitic limestone.
- 50–90% dolomite = calcitic dolomite.
- 90–100% dolomite = dolomite.

Dolostones are used locally for building, with a famous example being the Magnesian Limestone (Cadeby Formation, Upper Permian) of Yorkshire, England. Petrographically, the Magnesian Limestone shows a wide range of textures. Figure **70** shows 'Hampole', a dolomitic limestone with the bioclastic texture of the original limestone, still partially preserved despite being dolomitized.

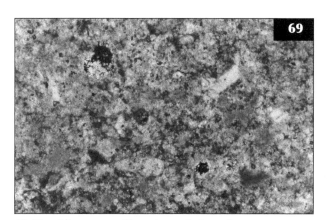

**68** Mokattam Formation limestone (Egypt) consisting of Nummulities bioclasts (white) cemented by microcrystalline calcite (micrite, brown). From the core of Khafre's Pyramid at Giza; XPT, ×25.

**69** 'Tura' limestone (Egypt) consisting of sparse microfossil shell fragments (pink) set in a matrix microcrystalline calcite (micrite, brown). From the casing of Khafre's Pyramid at Giza; XPT, ×150.

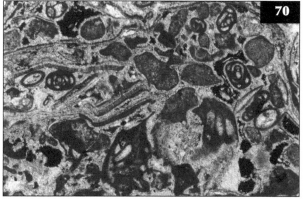

**70** 'Hampole' dolomitic limestone (Yorkshire, England) consisting of bryozoans, coral, and shell fragment bioclasts (brown) with a mantle of dolomite crystals (pink). Macropores are shown yellow/green; XPT, ×35.

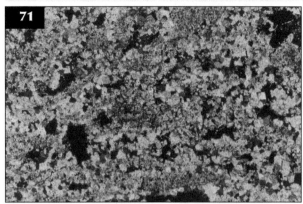

**71** 'Huddleston' dolomitic limestone (Yorkshire, England) consisting mainly of dolomite crystals (pink). Macropores are shown black; XPT, ×35.

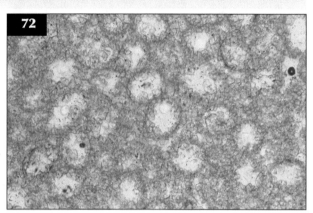

**72** 'Highmoor' dolomitic limestone (Yorkshire, England) consisting of dolomite crystals (grey) with a cellular structure. Macropores are shown yellow; PPT, ×150.

**73** Travertine (Turkey) consisting a calcite crystals (light brown) and including small irregular cavities (yellow); PPT, ×35.

**74** Travertine consisting of calcite (pink) with distinct banding; XPT, ×35.

Figure **71** shows 'Huddleston', which is a dolomitic limestone with a coarsely crystalline, porous texture. Figure **72** shows 'Highmoor', which has a highly porous, cellular texture that represents an original ooidal/peloidal fabric in which the cores of the spheroidal framework grains have been leached out.

Travertine is a porous but overall well consolidated calcitic limestone, which forms around springs. It is used for masonry locally, and exported for flooring and lining. It is typically whitish, yellow to brown, banded, and compact, with some characteristic irregular cavities. Figure **73** shows a sample of travertine that has been cut

parallel to the banding, while Figure **74** shows a different travertine sample that has been cut across the banding.

Onyx-marble is an exceptionally fine-grained, generally translucent and banded variety of calcite that forms at warm springs. It is used for decorative purposes. Figure **75** shows an onyx-marble that is translucent green with brown banding in hand specimen.

## OTHER SEDIMENTARY ROCKS

An ironstone (**76**) is a sedimentary rock that contains more than 15% iron, which may be present as variable proportions of iron-bearing minerals such as goethite,

75 Onyx-marble that has been stained in accordance with Dickson's method. The red bands consist of nonferroan calcite while the blue bands consist of ferroan calcite. The white (unstained) bands are composed of dolomite; PPT, ×25.

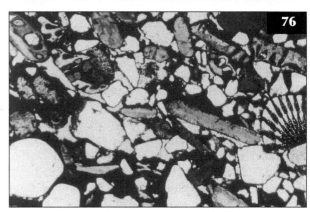

76 Ironstone (England) consisting of shell fragments (light brown) and quartz grains (white), cemented by iron-bearing minerals (black); PPT, ×25.

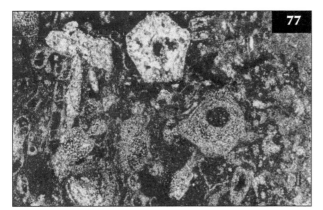

77 Flint from the Chalk (Upper Cretaceous, England) consisting of microcrystalline and cryptocrystalline silica (grey). Fossil remnants representing the rock texture prior to chertification are seen in this example; XPT, ×35.

78 Alabaster consisting wholly of gypsum crystals; XPT, ×35.

siderite, and berthierine. Sedimentary ironstones may be hard and calcareous and contain quartz grains, ooliths, pebbles, and a wide variety of macrofossils. They are used for building locally and are characterized by their strong yellow–brown or orange–brown colours when weathered.

Flints (77) are composed of pure, dense, crypto-crystalline silica (chert). They occur as layers of black nodules with a white outer cortex, throughout much of the Chalk of western Europe or in gravels weathered from the Chalk. They are used for building in areas where there are few other suitable stones. They may be used as rubble wall fill or as squared and knapped flint wall facings.

Alabaster (78) is a decorative stone composed of gypsum ($CaSO_4.2H_2O$) in a fine-grained, massive, and compact form. The purest form of alabaster is white and translucent, but traces of iron minerals produce light brown, orange, or red streaks. The term alabaster is sometimes incorrectly used to market decorative stones composed of calcium carbonate such as onyx-marble. True alabaster is composed of gypsum.

## STONE FROM METAMORPHIC ROCKS

All rock types may be transformed into metamorphic rocks long after the original time of their formation. Metamorphism results from alteration by heat and pressure either from local metamorphism at contact aureoles and fault zones, or from regional metamorphism in mountain belts. The mineral fabric of the pre-existing rocks may be reformed to give larger crystals of the original minerals, or form new minerals that are stable in the newly imposed environment. The overall chemistry (elemental composition) of the rock remains the same, unless new chemicals were introduced.

The classification of metamorphic rocks depends on the nature of the original rock and the degree to which it was metamorphosed. At moderate temperatures, low-grade metamorphic rocks are formed. These are very fine-grained rocks such as slate. At higher temperatures, the rates of chemical change increase and new minerals are able to grow to larger sizes. This forms medium-grade metamorphic rocks such as schist, metaquartzite, and marble. At the highest temperatures, the original fabric is obliterated to form coarse-grained rock that is foliated and consists of minerals that are stable at high temperatures. Gneiss is the main high-grade metamorphic rock type. A classification of metamorphic rocks for engineering purposes is provided in *Table 10*.

**Table 10** Geological classification of metamorphic rock for engineering purposes (a modified form of this table appears in BS 5930:1999)

| Foliated rocks | | Nonfoliated rocks | Grain size (mm) |
|---|---|---|---|
| **Grain size description** | | | >20 |
| Coarse | **Gneiss** Well developed but widely spaced foliation sometimes with schistose bands | **Marble** **Metaquartzite** **Granulite** **Hornfels** **Amphibolite** **Serpentinite** | 2–20 |
| | **Migmatite** Irregularly foliated: mixed schists and gneisses | | |
| Medium | **Schist** Well developed undulose foliation; generally much mica | | 0.06–2 |
| Fine | **Phyllite** Slightly undulose foliation; sometimes 'spotted' | | 0.002–0.06 |
| | **Slate** Well developed plane cleavage (foliation) | | <0.002 |
| Glassy | **Mylonite** Found in fault zones, mainly in igneous and metamorphic areas | | Amorphous or cryptocrystalline |

Schist is a common metamorphic rock that exhibits an alignment of platy and other elongated minerals called schistosity. As schist will part relatively easily along the plane of schistosity it is used locally for building. Figure **79** shows a garnet-mica schist from Norway. Gniess is a foliated rock with minerals aligned alternately in fairly large-scale parallel layers known as gneissose banding. This banding produces some attractive stones. Figure **80** shows a charnokite gneiss from India that is used for flooring and lining.

Metaquartzite is produced when a sandstone is metamorphosed causing the constituents to recrystallize into an interlocking mosaic of quartz. Figure **81** shows

'Alta Quartzite', a silver grey–green micaceous quartzite from Norway that is used for flooring and paving. Figure **82** shows 'Azul de Macaubas', a decorative blue quartzite from Brazil.

Marble is a prestigious stone formed by the metamorphism of limestone. The limestone is recrystallized to produce an interlocking granular mosaic of roughly equal-sized calcite crystals. Pure marble is made entirely of calcium carbonate and is white in colour and thin slabs may appear translucent. Impurities in the original limestone produce coloured marble (black, grey, green, pink, red, yellow) that may have a mottled or veined appearance. 'Statuario Venarto' is a Carrara

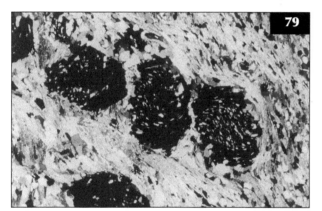

79 'Furuland Schist' (Norway) consisting of well aligned crystals of quartz (grey/white), muscovite mica (orange/pink), and garnet (black); XPT, ×35.

80 'Shivakashi' charnokite gneiss (India) with a foliated texture, consisting of quartz (grey) and feldspar (grey, twinned) with minor garnet (black) and orthopyroxene (orange); XPT, ×35.

81 'Alta Quartzite' (Norway) consisting mainly of quartz (grey) with minor proportions of feldspar and muscovite mica (pink); XPT, ×25.

82 'Azul de Macaubas' (Brazil) consisting mainly of quartz (grey) with minor proportions of the fibrous mineral dumortierite (brightly coloured); XPT, ×35.

marble, which is white with grey veining. Figure **83** shows a white area of 'Statuario Venarto' and Figure **84** shows an area that includes a grey vein. Figure **85** shows 'Sky Blue', a pale blue marble. Serpentine marble (ophicalcite) contains sufficient quanities of the mineral serpentine to change its colour, usually to green. Figure **86** shows a mottled grey/black/green marble from China. Figure **87** shows 'Indian Green', a dark green serpentine marble from India.

**83** 'Statuario Venarto' marble (Italy) consisting of fine white, interlocking rhomboidal calcite crystals; XPT, ×35.

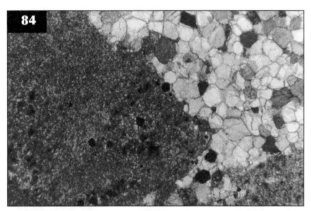

**84** 'Statuario Venarto' marble (Italy) consisting mainly of fine white calcite crystals (upper right) with a grey vein of very finely crystalline dolomite (left); XPT, ×150.

**85** 'Sky Blue' marble consisting chiefly of calcite with minor proportions of clinopyroxene (bright pink/blue/green) and olivine (upper left, orange); XPT, ×35.

**86** A mottled grey/black/green marble (China) consisting mainly of calcite (light brown) and serpentine (grey) with minor muscovite mica (pink/yellow/green); XPT, ×35.

**87** 'Indian Green' marble (India) consisting chiefly of calcite (pink) with minor serpentine (blue) and tremolite (orange); XPT, ×150.

# PETROGRAPHY OF STONE DEFECTS AND DECAY

Despite its image as an eternal material, even good-quality building stone has a limited life. Deterioration and decay may result from inherent defects within the stone, or more usually, from the processes of weathering or attack from pollution. The causes and degree of stone defects and decay are eminently suited to assessment by petrographic examination.

Inherent flaws found within stones include fractures and shakes, which act as planes of weakness and may cause stone unit failure. Stone containing these features would normally be discarded. Other natural stone features such as bedding planes, joints (see **32**) or stylolites, also have the potential to act as planes of weakness. These can be assessed in thin section to determine the open width, the degree of wall contact, and the types/stability of any filling material. Figure **88** shows 'Filetto Rosso' from Italy, which is a hard, polishable, cream-coloured limestone (sparse biomicrite) with red/pink stylolites. Stylolites are irregular suture-like boundaries, generally running independently of the bedding, which are present in some limestones. Figure **89** shows 'Vtratza' limestone from Bulgaria, which is a compact, cream-coloured limestone (biomicrite), which sometimes includes fossil burrows that are lined with organic matter. These appear like knots in wood and these 'knots' can fall out of the stone, leaving holes.

The most important stone decay mechanisms are salt crystallization, attack by acid gases in the air, and frost action (Ingham, 2005a). Salt crystallization occurs when a solution of salt/s in water is deposited under drying conditions on the surface of the masonry and/or within its pores. The growth of the salt crystals is an expansive process that causes powdering, scaling, and delamination of the outer stone. Pollution from industry and transport causes masonry decay by attack from acid gases in the air. Sulfur (and nitrogen-based) gases react with moisture to produce acidic solutions that can directly attack calcareous stones. Exposed masonry surfaces are dissolved away and salt crusts (often including particulates and other dirt) are deposited on more sheltered surfaces, giving a soiled appearance (**90**).

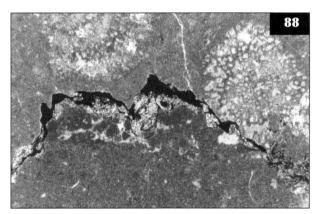

**88** 'Filetto Rosso' limestone (Italy) with a stylolite that is partially open (black) and partially filled with clay and iron minerals (orange); XPT, ×35.

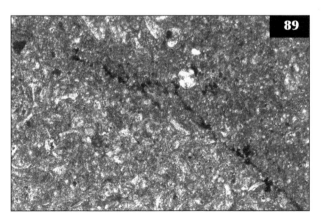

**89** View of 'Vtratza' limestone (Bulgaria) showing organic matter (dark brown) lining a fossil burrow. The sample has been stained in accordance with Dickson's method; PPT, ×150.

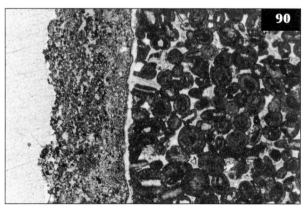

**90** Pollution crust consisting of gypsum and soot (left) on a Portland limestone surface. Gypsum (white) also fills the macropores near to the outer surface; PPT, ×35.

On areas of buildings that can remain wet, frost action can cause cracking and surface scaling (**91, 92**). The pore structure of stone plays an important part in determining its resistance to frost, with impermeable and free-draining stones being more durable. Polarizing and fluorescence microscopy are used to determine the pore structure and crack patterns of stones with a view to determining the degree of damage attributable to freeze–thaw mechanisms and prediction of future performance (Ingham, 2005b).

Certain organisms that colonize masonry have the potential to cause biodeterioration. Trees, creepers, and mosses (**93**) cause physical damage from their roots, while certain bacteria, fungi, and lichen deposit corrosive organic acids. Optical microscopy is a powerful tool for assessing bioreceptivity and biodeterioration of colonized building stone, allowing assessment of biofilm characteristics in relation to their substrate. Successful applications include locating and evaluating biologically-induced physicochemical changes, such as rhizoid-induced microcracking, lichen-induced pitting or perforating, and bacterially triggered acid leaching (Dreesen *et al.*, 1999).

A problem particular to certain marbles is nonreversible expansion caused by thermal cycling. For thin cladding panels this can cause bowing or warping of the panels on buildings, which could potentially lead to panel/fixing failure. The problem seems to be closely related to the

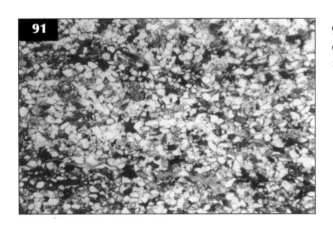

**91** Frost damage of a sandstone window sill with cracks (blue) running along the bedding planes; PPT, ×35.

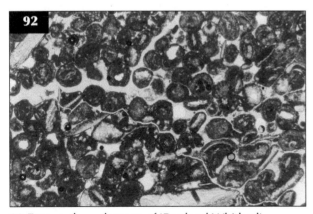

**92** Freeze–thaw damage of 'Portland Whitbed' limestone with fine cracks (yellow) running around ooliths and breaking the bonds between them; PPT, ×35.

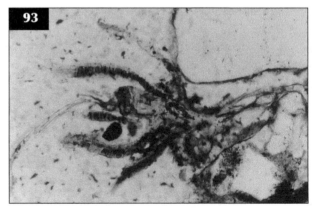

**93** Moss colonizing a 100-year-old masonry surface; PPT, ×300.

microstructure of the marble. Marble with xenoblastic texture seems to to be less prone to bowing than marble with a granoblastic texture. Xenoblastic marble has irregular grain boundaries that interlock with each other. Granoblastic marble has crystals that are almost an equal width in all directions. The grains are typically polygonal with almost straight boundaries that often meet at triple points. The microtexture of marble can be readily studied using the polarizing microscope. The definition of the crystal boundaries can be enhanced by using fluorescence microscopy or, alternatively, by etching (Alm *et al.*, 2005). Figure **94** shows a marble with granoblastic texture from a cladding panel that was suffering from bowing and exhibited a deflection of 15 mm.

Fire damage is a cause of stone decay that can be assessed petrographically to determine the heating history and depths of damage (Ingham, 2007). The changes caused by heating of natural stone include spalling, cracking (**95**), discolouration, nonreversible expansion, and calcination. Red discolouration caused by oxidation of iron compounds commences at around 300°C (**96**). Importantly, this corresponds with the onset of significant strength loss and it can be used to detect the depth of the 300°C thermal contour through the stone section. Care must be taken to ensure that the pink/red colour is attributable to heating, as some stones are naturally pink or red. In addition, some types of natural stone product are deliberately heat-treated by the stone

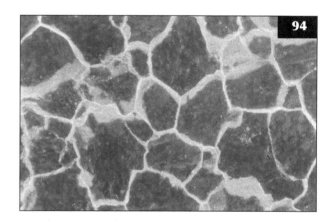

**94** Close view of a marble from a bowing cladding panel using fluorescence microscopy, showing granoblastic texture with calcite crystals appearing dark green and intercrystalline space shown as light green; UV, ×150.

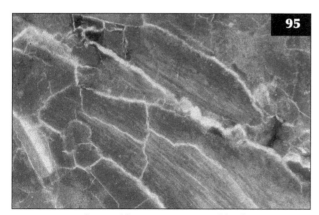

**95** Microcracking of limestone caused by heating near to the surface of fire-damaged masonry block; UV, ×150.

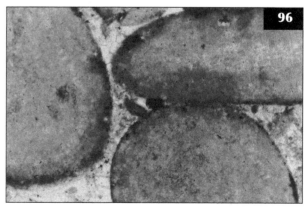

**96** Red discolouration of limestone from a fire-damaged masonry block; PPT, ×150.

producer to change the colour in order to increase sales. Figure **97** shows a microscopical view of sandstone that was deliberately heat-treated to change its colour from light brown to a more marketable deep red. Other important heat-induced changes in stone include cracking or shattering of quartz resulting from the α- to β- quartz phase transition at 573°C and the calcination of limestone and marble at 800–1000°C. The main heat-induced changes that affect different stone types are summarized in *Table 11*.

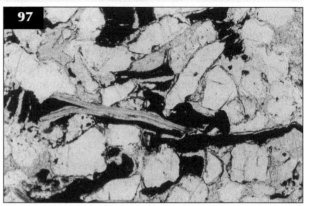

**97** Sandstone that has been heat-treated to redden its appearance. Iron-bearing minerals have been oxidized (black). Quartz grains appear white and pore spaces are shown yellow; PPT, ×150.

**Table 11** Changes caused by heating of various types of natural stone that may be observed either visually or microscopically (Ingham, 2007)

| Heating temperature °C | Stone type | | | |
|---|---|---|---|---|
| | Limestone | Sandstone | Marble | Granite |
| 250 | Pink or reddish-brown discolouration starts at 250–300°C but may not become visible until 400°C | Red discolouration starts at 250–300°C but may not become visible until 400°C | Heating marble through a range of temperatures causes nonreversible expansion known as thermal hysteresis | At less than 573°C, if heating rate is <1°C per minute the thermal expansion is fully reversible. If heating rate is >5°C per minute the expansion is not totally reversible |
| 300 | | | | |
| 400 | Discolouration becomes more reddish at 400°C | | | |
| 600 | Calcination of calcium carbonate commences at 600°C | Heating above 573°C causes internal rupturing of quartz grains with associated weakening and friability. Clay minerals in the cement disintegrate (kaolinite up to 600°C, chlorite above 600°C) | Above 600°C complete disruption due to differential expansion, becomes friable and reduces to powder | Develops cracks or shatters at 573°C due to quartz expansion |
| 800 | Calcium carbonate calcines to a grey–white powder at 800–1000°C with associated loss of strength | Red discolouration may persist until 1000°C. Any calcium carbonate cement calcines to powder at 800–1000°C causing disintegration | | Differential thermal expansions at higher temperatures (900°C) gives rise to tensile and compressive stresses causing permanent strain in the stone |
| 1000+ | Melting starts | Melting starts | Melting starts | Melting starts |

# Roofing slate

## INTRODUCTION

Slate is a rock derived from fine-grained sediments (mudstones or siltstones, or in some cases volcanic tuffs), which have undergone low-grade metamorphism and deformation as results of tectonic compression. Slate comprises a high proportion of platy phyllosilicate minerals that align during deformation, imparting a cleavage to the rock along which it can be split into thin sheets (slaty cleavage).

The ability to split slate along planes of cleavage makes it suitable for use as a building material, in particular a roofing material. Slate and other fissile rocks have a long history of use in roofing and good roofing slates have been known to survive on buildings for hundreds of years.

Selected commercial applications of petrographic examination to roofing slate investigation include:

- Initial suitability assessment of slate from new sources.
- Routine quality assurance of slate from the factory production run.
- Confirming the identity/quality of slate for purchasers.
- Diagnosing the causes of in-service deterioration/failure.
- Matching slate types for restoration of historic buildings.

## TESTING ROOFING SLATES

As with other natural stone resources, slate is variable in character and testing is required as part of the quality assurance for slate products. Testing requirements for roofing slate are usually specified either by the EN or the ASTM.

In Europe, to ensure that roofing slate is fit for purpose, the CEN requires the manufacturer to undertake a programme of initial type testing to evaluate the composition, strength, and potential durability of every new source (British Standards Institution, 2004). Once it has been established that the material is suitable for roofing applications, further testing of the factory product is required at regular intervals for quality assurance. Further testing is sometimes required by the manufacturer, the purchaser, or third parties to investigate nonconformities/complaints, confirm identity, or to diagnose the causes of in-service failure.

It is a EN 12326-1 (British Standards Institution, 2004) requirement that petrographic examination be performed on slates from new sources as part of initial type testing and at least annually (or every 25,000 tonnes if sooner) for existing sources, as part of factory production control. Additionally, petrography is identified as a standard tool for the resolution of quality disputes. Petrographic examination should be the first test carried out to check that the rock is actually slate and decide whether the product falls within the scope of EN 12326-1 classification.

The standard suite of EN tests also includes assessment of slate dimensions, bending strength, water absorption, carbonate and noncarbonate carbon content, along with a sulfur dioxide exposure test, and a thermal cycling test. A full programme of examination, analysis, and testing for factory production quality control requires assessment of up to 100 whole slates. The slates for assessment are selected at random from the production run. When investigating in-service failure, a number of slates would typically be removed from the roof. These would usually be chosen to represent the range of slate quality present, including both apparently defective slates and apparently nondefective comparison samples.

In the USA, roofing slate is specified in accordance with ASTM C406 (ASTM International. 2006), which covers the material's characteristics, physical requirements, and sampling requirements. The standard requires that modulus of rupture, absorption, and depth of softening tests are conducted on a set of representative slates. Petrography is not currently a requirement of the standard.

## PETROGRAPHIC EXAMINATION AND COMPLEMENTARY TECHNIQUES

Within the European Union, petrographic examination of roofing slate is performed following a procedure given in EN 12326-2 (British Standards Institution, 2000). Petrographic examination comprises a combination of visual and low-power examination in hand specimen and more detailed high-power microscopical examination in thin section and polished specimen, which may be supplemented by mineralogical analysis by XRD.

For initial type testing or factory production control, petrographic examination is conducted on a single slate randomly selected from the assessment batch. Following arrival in the laboratory, the sample is first examined in the as-received condition using the unaided eye and low-power stereo-zoom microscope at magnifications of up to ×60. Slate features observed at this stage include colour, thickness, 'grain', relict bedding, open and incipient cracks, joints and veins, nodules and inclusions, and surface discolouration. The initial examination is used to determine the most appropriate location for thin sections and polished specimens to be taken for further, more detailed high-power microscopical examination. Thin sections are prepared from slices representing three different orientations, comprising two perpendicular to the cleavage (at right angles to each other) and one parallel to the cleavage. Polished specimens are normally prepared to enable reflected light examination of opaque minerals.

Slates are very fine-grained and relatively difficult to resolve by optical microscopy, so it is desirable to enhance the optical observations by SEM and associated microanalysis. Mineralogical analysis by XRD is particularly useful to determine the minerals that are present, but, for routine investigations, only gives semiquantitative estimates of proportions. The accuracy of a quantitative determination for a particular mineral (e.g. quartz) can be improved by adding a known amount of that mineral ('spiking') to the sample (Walsh, 2002). Relatively accurate elemental analyses can be obtained by XRF analysis.

## PROPERTIES OF ROOFING SLATES

### MINERALOGY
The essential minerals of typical slate are quartz and phyllosilicates (white micas and chlorite) with various accessory minerals (**98**). The accessory minerals can include feldspars, carbonates (calcite, dolomite, magnesite, siderite), iron oxides (haematite, magnetite, limonite), iron sulfides (pyrite, marcasite, pyrrhotite), other ore minerals (rutile, anatase, ilmenite), and miscellaneous minerals (graphite, apatite, chloritoid, tourmaline, epidote, and zircon).

Slate properties are largely dependent on mineralogy. The proportion of platy minerals controls splitting, quartz imparts hardness and durability, while the presence of some accessory minerals adversely affects durability (Walsh, 2003). The colour of a slate is also determined by its mineralogy. The most common colour is grey which is attributed to a combination of quartz with a small amount of graphite (carbon). Increased amounts of graphite darken the colour to black, a colour also associated with pyrite, which, if finely disseminated, tends to darken slate colour. Iron oxides such as haematite (**99**) give red and purple colours, while green slates contain more chlorite (**100**) and occasionally epidote. Some slates exhibit round/elliptical bleached spots called reduction spots. These are formed around a particle of organic matter, which acted as a reducing agent and converted purple ferric oxide to green/yellow ferrous oxide (**101**).

As slate is formed, by diagenetic and metamorphic processes, the proportions and types of minerals present change. New minerals grow as clay minerals are lost, the chemical composition and atomic structure of phyllosilicates change, crystallinity increases, and the grain size increases (Walsh, 2002). An understanding of mineralogy gained from microscopical examination aids the interpretation of metamorphic grade, which in turn gives an indication of the quality of slate as a roofing material.

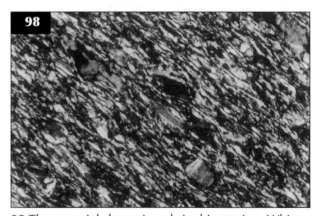

**98** The essential slate minerals in thin section. White mica (white/grey/yellow), quartz (grey), and lenses of chlorite (blue). 'Villar Del Rey' slate from Spain; XPT, ×300.

As slate is very fine-grained, optical resolution of minerals is difficult and consequently, definitive determination of mineralogy usually requires combination of several different microscopical and analytical methods. The approximate quantity of each mineral present may be estimated visually through the microscope or by point counting to give a volume percentage. A more accurate approach involves a combination of optical microscopy, SEM, elemental analysis by EPM or XRF, and mineralogical analysis by XRD.

## CLEAVAGE

Slate is found in mountainous areas where tectonic forces have compressed the rocks by folding. In response, minerals grow perpendicular to the direction of maximum stress producing slaty cleavage (102). Cleavage forms independently of sedimentary bedding (along which some sedimentary rocks can be split) and is often at high angle to it. The type and intensity of cleavage controls every aspect of slate production, from ease of extraction to the characteristics of the finished product. At a microscopical level, cleavage planes can be

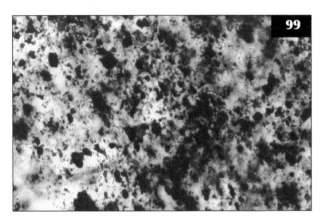

**99** Close view of purple coloured slate in thin section showing the haematite (red/black) that imparts its colour. 'Penrhyn' slate (Wales); PPT, ×600.

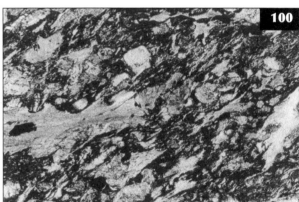

**100** 'Westmorland Green' slate (Cumbria, England) formed from deposits of volcanic ash and containing chlorite (green), which provides its green colour; XPT, ×150.

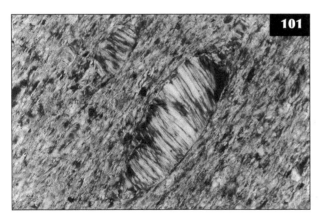

**101** Close view of a green reduction spot in 'Penrhyn' slate (Wales), showing a concentration of chlorite (grey) that gives its colour; XPT, ×300.

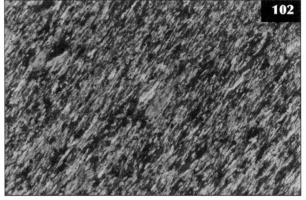

**102** Slate with a continuous, well developed cleavage (running diagonally in the field of view). 'Cwt-Y-Bugail' slate (Wales); XPT, ×300.

seen as closely spaced, subparallel dark seams of phyllosilicates and graphitic material along which the rock can be split (**103**). A cleaved rock is divided into two domains: cleavage domains consisting of strongly aligned minerals, which are separated by microlithons where the minerals may or may not be aligned (Walsh, 2002). Microscopical examination of the slate domains and classification in terms of their shapes, spacing, proportion, and fabric allows an understanding of the factors controlling the thickness and texture of the slate and its metamorphic grade. In general, with increasing metamorphic grade, the spacing between the cleavage domains decreases, the proportion of rock occupied by cleavage domains increases, and the minerals in the microlithons become more aligned (Walsh, 2003).

Crenulation cleavage consists of microfolds found on the cleavage surface, owing to the development of cleavage in a rock with a pre-existing fabric (**104**). If well developed, it is an undesirable characteristic in slates that are being quarried because it may cause the slate to break too readily in a direction other than that of the primary cleavage.

## GRAIN

The grain is the secondary splitting direction of the slate and is usually parallel to the direction of maximum tectonic stress. This secondary plane of weakness is recognized by quarrymen and exploited in the winning of slate as the pillaring line. Grain is caused by some of the platy minerals being aligned with their flat face parallel to and/or elongated along the direction of maximum stress during deformation. The grain may be visible as a series of fine lineations on the cleavage surface, but can often only be determined by microscopical methods. This is particularly true if the grain direction is nearly coincident with the primary cleavage and makes a low angle with the cleavage surface. In respect of roofing slate, the grain should ideally run normal to the cleavage direction (along the length of the slate) so that if the slate breaks under load on the roof, the pieces will still be held in place by the fixing nails.

### RELICT SEDIMENTARY FEATURES

The development of cleavage has often not obliterated all signs of the original sedimentary bedding. Bedding can be indicated by the occurrence of small-scale banding of coarser and finer grain size and differing mineral composition, running at an angle to the cleavage (**105**). Fossils are rarely found in slates as the development of cleavage usually results in the

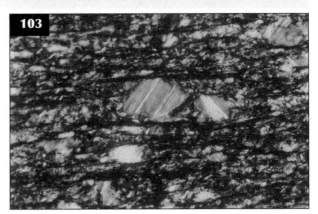

**103** Microfabric of slaty cleavage seen in thin section. Narrow, planar cleavage domains (black) pass between microlithons including chlorite lenses (blue). 'Aberllefenni' slate (Wales); XPT, ×300.

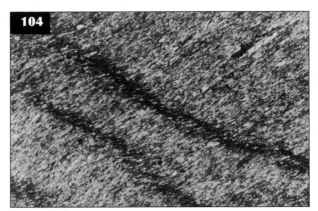

**104** Slate with crenulation cleavage, seen in thin section. 'Dequesa' slate (Spain); XPT, ×150.

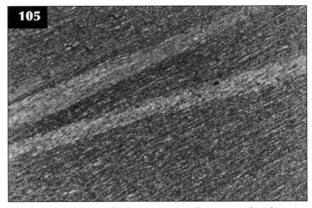

**105** Section through the primary cleavage of a slate with relict bedding running at approximately 10°. 'Cwt-Y-Bugail' slate (Wales); PPT, ×35.

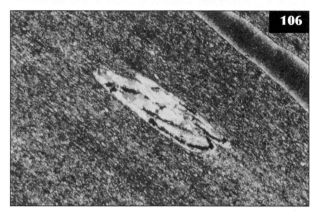

**106** Fossil remnant in a Spanish slate. Composed of calcite (pink) and indicative of incomplete metamorphism; XPT, ×35.

eradication of fossils or renders them too obscure to be recognized. Fossil remnants are sometimes observed within slates that were incompletely metamorphosed and where the slaty cleavage is poorly developed (106).

## JOINTS AND VEINS

Slate is unloaded as it is brought to the Earth's surface and the rock fractures in a set of directions other than the main cleavage; such fractures are known as joints. These joints can be exploited to extract the slate blocks. However, sometimes the joints are tight (with no visible break) and slates may be produced that include these joints, which may act as planes of weakness. In geological time, joints can open slightly and become tightly filled by minerals forming a vein. The in-service performance of veins depends on the minerals they contain and how completely they have infilled the joint and cemented the joint walls. Vein-filling minerals can include quartz (107), carbonates, iron sulfides, and chlorite (108). Rapid decay of unsound vein materials may lead to splitting and discolouration of the slate (109).

## INCLUSIONS AND NODULES

Many slates exhibit some small irregular concentrations of their constituent minerals that are seen to protrude from the cleavage surface as nodules. An abundance of nodules making the cleavage surface uneven may cause the slate not to lie flat on the roof and compromise the weatherproofing. Mineral concentrations not protruding

**107** A slate from Argentina with a quartz vein (grey); XPT, ×35.

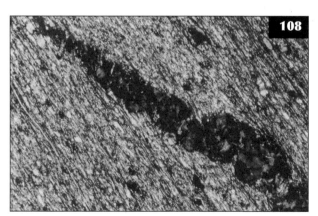

**108** A chlorite vein (blue) in 'Castilla' slate (Spain); XPT, ×150.

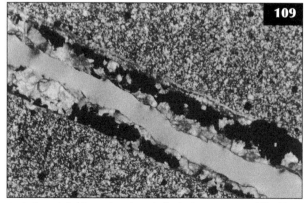

**109** Spanish slate that has failed along a vein filled by calcite (brown) and pyrite (black); XPT, ×35.

from the surface may be referred to as inclusions. The inclusions may be dispersed or concentrated in bands along the grain. Common mineral inclusions include quartz, carbonates, and iron sulfides, and different minerals may occur together (110).

Of particular importance are several iron sulfide minerals that can be present in slate, the most common one being pyrite. Pyrite occurs in a number of morphologies including large well formed cubic crystals (111), which are usually relatively stable. Other pyrite morphologies comprise smaller and less crystalline rounded grains (112) that are often metastable. Less common iron sulfide minerals include marcasite (a polymorph of pyrite) and pyrrhotite.

# WEATHERING AND DETERIORATION OF ROOFING SLATE

The natural weathering of rock masses usually takes place very slowly and leads to the breakdown and eventual decomposition of rock at the Earth's surface. Slate quarried for roofing would normally be extracted from fresh rock beneath the weaker weathered layer. Following production, roofing slates are placed on the most exposed part of buildings and can be subjected to severe weather conditions. To perform satisfactorily, slates must resist physical deterioration from driving rain, high winds, and extremes of temperature. Minerals within slates are also subjected to chemical weathering by a variety of reactions including ionization, hydrolysis, and oxidation. The most significant chemical weathering reactions are oxidation of iron sulfides and the reactions of carbonate minerals with acids from rainwater and polluted atmospheres. The oxidation of pyrite comprises a series of reactions, some of which involve bacteria (e.g. *Thiobacillis ferrooxidans*) to produce iron oxides (rust) and sulfuric acid. The disintegration process of pyrite may be simplified as follows (Blanchard & Sims, 2007):

$$FeS_2 + 9H_2O \rightarrow Fe(OH)_2 + 2H_2SO_4 + 7H_2$$

Some forms of iron sulfide are more prone to weathering than others. Pyrrhotite and marcasite are the most reactive minerals and less crystalline forms of pyrite with

**110** Spanish slate with a band of inclusions running through the full thickness of the slate. The band runs along the grain and the inclusions comprise pyrite (black), quartz (white/grey), and calcite (pink); XPT, ×35.

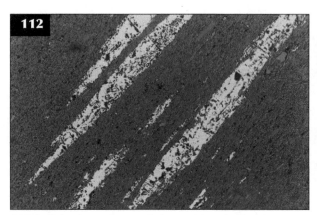

**111** Polished specimen of slate with cubic crystals of pyrite (yellow); PPR, ×150.

**112** Polished specimen of slate with anhedral lenses of pyrite (yellow); XPR, ×75.

large surface areas are more susceptible than large cubic crystals (Evangelou, 1995). The rusting of iron sulfide inclusions causes delamination and other deterioration owing to expansion caused by the formation of iron oxides (113). Rust deposits from pyrite often discolour the cleavage surface (114).

Some slates contain significant quantities of carbonate minerals, which may occur as discrete inclusions (see 106), concentrated in veins (see 109) or bands (see 110), or along bedding planes (115). Some rocks sold as slate have been found to be metamorphosed marls, composed almost entirely of dolomite and calcite (116). Carbonate minerals are particularly susceptible to attack from acids present in polluted atmospheres (117) and sulfuric acid

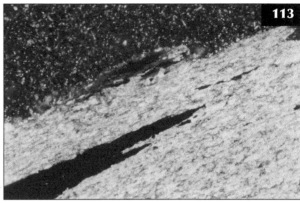

**113** Weathering of pyrite inclusions (black) within slate. The inclusion at the upper surface exhibits brown rust deposits; XPT, ×150.

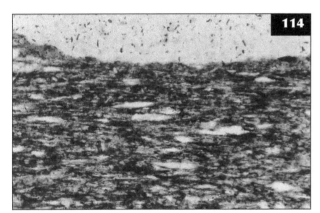

**114** Discolouration of the upper slate surface. Rust deposits from pyrite oxidation form a thin brown layer; PPT, ×150.

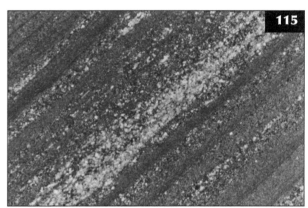

**115** Calcite (pink) concentrated along a bedding lamination. Brazilian mudstone/siltstone being sold as roofing slate; XPT, ×35.

**116** Metamorphosed marl chiefly comprising carbonate minerals (mainly dolomite). This was imported from China as roofing slate; XPT, ×300.

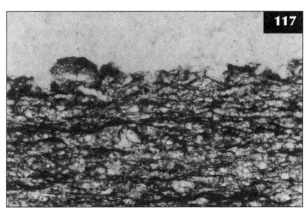

**117** Deterioration of carbonate minerals at the upper surface of metamorphosed marl 'roofing slate' from China; PPT, ×300.

**118** Weathering delamination in Cornish slate after 70 years on a roof. The delaminations are highlighted by the yellow resin used to impregnate the specimen; PPT, ×75.

**119** Pre-existing delamination (yellow) of Brazilian mudstone/siltstone being sold as roofing slate; XPT, ×150.

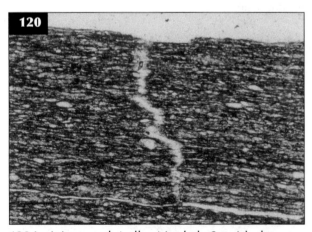

**120** Incipient crack (yellow) in shaly Spanish slate. The crack intersects with a delamination; PPT, ×75.

released when pyrite decays. Calcite reacts with sulfuric acid to form the expansive mineral gypsum (calcium sulfate hydrate) as follows:

$$CaCO_3 + H_2SO_4 + H_2O \rightarrow CaSO_4.H_2O + CO_2$$

Where inclusions of reactive pyrite and calcite occur together they can cause rapid deterioration. Where these inclusions run through the full thickness of the slate the deterioration can cause the slate to split. EN 12326-1 (British Standards Institution, 2004) recognizes that certain slates may contain acid-soluble carbonate minerals. The standard permits the use of certain slates containing carbonate minerals providing that they meet defined strength and durability criteria and that they are sufficiently thick to compensate for potentially increased weathering rates.

The action of weathering on roofing slates, often involving a combination of physical, chemical, and biological mechanisms, can result in delamination, cracking, and splitting. Delamination involves cracking parallel to the cleavage surfaces that starts from the exposed dressed edge and moves inwards, towards the centre of the slate (**118**). Incompletely metamorphosed slates of shaly character can be unacceptably weak and have been observed to exhibit delamination even prior to use (**119**). Delaminated or otherwise weakened slates are subsequently vulnerable to cracking or splitting perpendicular to the cleavage, or along the grain. Incipient cracks, where the slate is cracked but still holds together, may be visible on the cleavage surface as white lines and microscopically these cracks are sometimes seen to intersect with delaminations (**120**).

## INTERPRETATION OF SLATE TEST RESULTS

The first question to be answered is 'is the rock a true slate?' EN 12326-1 requires that roofing slate must be a true slate in the geological sense (with phyllosilicate minerals predominating and exhibiting a prominent slaty cleavage), otherwise the material does not fall within the scope of the standard and cannot be sold as complying with it. This geological classification can only be determined by petrographic examination. Petrography is also the first line of defence to check that material supplied is the slate type ordered and to detect if an inferior product has been substituted.

The petrographic examination will also determine if the slate has any inherent physical or petrographic defects or damage that would be detrimental to performance. EN 12326-1 requires that slate shall be free from such defects.

Once the roofing material has been established as being slate it should conform to the following physical and chemical test criteria in accordance with EN 12326-1:

- Individual slate thickness not less than 2 mm.
- Variation of thickness of individual slates shall not be greater than ±35% of the nominal thickness.
- Packed thickness of 100 slates shall not deviate by more than ±15% of the nominal thickness.
- The batch of slates shall have consistent length, width, edge straightness, rectangularity, and flatness in accordance with the requirements of the standard.
- Water absorption of ≤0.6%. Alternatively >0.6% but must pass a freeze–thaw resistance test.
- Pass thermal cycling test without exhibiting exfoliation, splitting, or other major structural changes.
- Noncarbonate carbon content of ≤2%.

Bending strength, thermal cycling, and sulfur dioxide exposure test results provide classifications that are used to calculate the minimum thickness that the slates must be to perform satisfactorily on a roof. Petrographic examination can be applied to failed test specimens to investigate the causes, extent, and significance of failure during thermal cycling and sulfur dioxide exposure tests (121).

The pass/fail criteria to the EN 12326-1 roofing slate standard are not very demanding and a pass does not necessarily guarantee the highest quality of slate. In practice the pass range includes a variety of different slate qualities and the standard gives no detailed guidance on quality classification, or any way of predicting the potential service life of roofing slate. To aid roofing slate quality classification, Hunt (2005) has proposed a scheme, which uses the results of the EN 12326-2 tests to give an indication of possible roofing slate service life (*Table 12*).

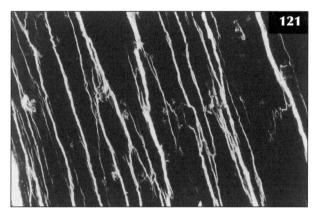

121 Delaminations (green) in a slate specimen from a sulfur dioxide exposure test, highlighted using fluorescent microscopy; UV, ×35.

**Table 12** Classification of roofing slate quality from results of EN 12326-2 tests (after Hunt, 2005)

| Criteria | Class A | Class B | Class C | Class D | Class E* |
|---|---|---|---|---|---|
| Flexural strength (MPa) | ≥70 | ≥60 | ≥50 | No limit | No limit |
| Characteristic modulus of rupture (MPa) | ≥55 | ≥45 | ≥35 | No limit | No limit |
| Water absorption (%) | ≤0.3 | ≤0.4 | ≤0.5 | ≤0.6 | ≤0.6 |
| Potential for pyrite oxidation | No | No | No | Yes | No |
| Carbonate content (%) | ≤3, preferably ≤1 | ≤3 | ≤5 | ≤1 | No limit |
| Conform to EN 12326-1 | Yes | Yes | Yes | Yes | Yes |
| Expected service life (trouble free) | >75 years | 40–75 years | 20–40 years | Up to 20 years | Up to 20 years, likely to exhibit variable colour change sooner |

\* Class E applies to 'roofing carbonate slates' as defined by EN 12326-1 ('rock used for roofing and external cladding, containing phyllosilicates and a minimum carbonate content of 20% and exhibiting a prominent slaty cleavage').

Most good-quality slates will change colour gradually as they slowly weather on a roof. Colour changes are noted during the thermal cycling and sulfur dioxide exposure test but colour stability is not part of the durability assessment criteria. However, such aesthetic changes are often of great concern to slate roof owners and if rapid unexpected changes occur, can cause durability concerns and dissatisfaction with the product. Slates containing a high proportion of carbonate minerals are susceptible to dramatic colour change (especially lightening in colour), that may also vary between individual slates to give a mottled roof appearance. Potential brown rust staining from pyrite oxidation can often be predicted by detecting the presence of iron sulfur minerals and understanding their reactivity potential.

In the USA, the ASTM C406 classification system applies. This uses the results of three physical tests to classify roofing slate into one of three grades of expected service life, $S_1$, $S_2$, or $S_3$ (*Table 13*). Although not a requirement of ASTM C406 it would be prudent to screen all roofing slate for potential defects by petrographic examination, in addition to conducting the required tests.

**Table 13** Classification of roofing slate quality from results of ASTM tests (ASTM C406)

| Classification | Minimum modulus of rupture across the grain ASTM C120 (MPa) | Maximum water absorption ASTM C121 (%) | Maximum depth of softening ASTM C217 (mm/inches) | Service life (years) |
|---|---|---|---|---|
| Grade $S_1$ | 62 | 0.25 | 0.05/0.002 | >75 |
| Grade $S_2$ | 62 | 0.36 | 0.20/0.008 | 40–75 |
| Grade $S_3$ | 62 | 0.45 | 0.36/0.014 | 20–40 |

# Aggregates

## INTRODUCTION

The term 'aggregate' covers a variety of materials used in the construction industry. Aggregates can be defined as 'particles of rock, manufactured or recycled material which, when brought together in a bound or unbound condition, form part or whole of an engineering or built structure'. The vast majority of aggregates used in construction work are for roads or concrete structures, where they are used unbound in the lower layers of road pavements, bitumen-bound in the upper road layers or bound by cement in concrete. Other applications for aggregates include mortar sand, railway ballast, and filter media. Each end use requires aggregate with specific properties in terms of particle size distribution (grading), shape and surface texture, strength, and resistance to degradation.

The principal applications of petrographic examination to investigation of aggregates are:

- Initial suitability assessment of aggregates from new sources.
- Routine quality assurance of aggregate from the production run.
- Assessing the potential for alkali–aggregate reactivity in concrete.
- Diagnosing the causes of in-service deterioration/failure.
- Matching aggregate types for restoration of mortar in historic buildings.

## PETROGRAPHIC EXAMINATION AND COMPLEMENTARY TECHNIQUES

Petrographic examination of aggregate is usually conducted following methods intended for the petrographic examination of concrete aggregates. In Europe, the standard for concrete aggregate (EN 12620) requires that petrographic description be conducted at least once every 3 years in accordance with the EN 932-3 method (British Standards Institution, 1997). EN 932-3 is a simplified method intended only for general classification of aggregate. However, a more thorough assessment is often required, for example, to determine potential alkali–silica reactivity and this would normally be conducted following BS 812: Part 104 (British Standards Institution, 1994). In America, ASTM C295 (ASTM International, 2008a) is the applicable standard and this method is often specified in the Middle East. An in-depth review of the different aggregate petrography methods used around the world is provided by Jensen (2007).

The simple EN 932-3 method involves examination of a relatively small number of aggregate particles in hand specimen, using a low-power microscope. The more thorough BS 812: part 104 and ASTM C295 methods involve hand separation of a large number of coarse aggregate particles, and/or point-counting of fine aggregate to determine the proportions of each aggregate constituent. In addition, for the more thorough methods, thin sections of aggregate particles are prepared for detailed high-power microscopical examination to confirm the hand specimen identifications. Aggregate particles are separated out into constituent groups based on differences in geological and engineering properties such as rock type, degree of alteration, weathering grade, particle shape, and surface texture.

If the microscopical examination proves insufficient to classify aggregate constituents, further investigation may be necessary. Mineralogical analysis by XRD is particularly useful for determining the composition of fines that are beyond the resolution of the optical microscope.

## AGGREGATE TYPE

The 'aggregate type' depends on the type of resource it is won from and the processing that the material undergoes to make the aggregate product. Aggregate type should be described as follows (adapted from Fookes *et al.*, 2001):

1. Whether natural or artificial.
2. If natural, whether crushed rock, gravel, or sand.
3. If a gravel or sand, whether uncrushed, partly crushed, or crushed.
4. If a gravel or sand, whether land won or marine.
5. If recycled, this should be stated.

Natural sands and gravels are superficial deposits of unconsolidated glacial, fluvial, or marine sediments. They are easily excavated by mechanical diggers and sieved or screened into different sizes. An important aggregates source, they are particularly valuable as concrete aggregate (see **154–156**) and mortar aggregate (see **279** and **280**). They are composed of the more durable rock fragments that have been released from their parent rock and abraded by physical weathering composition. For example, Figure **122** shows a beach sand composed of minerals weathered out from basalt and andesite lava flows on a Caribbean island. The properties of gravel, and to a lesser extent sand, largely depend on the rocks from which they are derived. Aggregates may be 'monomictic' in that they contain only one type of rock (e.g. **132**) or 'polymictic' with a number of different rock constituents (e.g. **131**). Certain natural deposits (e.g. wadi gravels) can be highly polymictic and may contain in excess of twenty different constituents (**123**).

Natural marine deposits are dredged from the sea floor and may be recognized by the presence of marine shell fragments (**124**) or remnants of encrusting marine organisms (**125**). Figure **126** shows a rare case of coral being used as coarse aggregate for concrete. On occasion, organic contaminants such as sand eels, may be found to be dredged-in with marine aggregate (French, 2005).

Crushed rock aggregates are produced from a variety of consolidated rocks that are at the surface, or near enough to the surface, for extraction to be economic. The rocks are quarried by blasting, before being crushed and screened into different particle sizes. Hard, compact types of limestone are major source of crushed rock aggregate (see **158**). Sandstones are a comparatively minor source because they tend to be friable and porous, except those that have been strongly cemented by silica to form sedimentary quartzites (see **155**). Igneous and metamorphic rocks provide a fair proportion of crushed rock aggregate, especially for roadstone (see **364**).

Artificial (or manufactured) aggregates are often derived from industrial waste. These include power station waste (pulverised fuel ash [PFA] and furnace bottom ash), blastfurnace slag, colliery spoil, china clay waste, slate waste, spent oil shale, steel slag, and incinerated refuse. They are increasingly being used as alternatives to natural aggregates as a way of conserving natural resources and disposing of waste arisings (Sherwood, 2001). Figure **127** shows sintered PFA which is used as lightweight concrete aggregate. Figure **128** shows boiler clinker aggregate, which is found in the concrete of historic 'filler-joist' floor construction. Figure **129** shows flint aggregate that has been heat-treated (calcined) to give it a white colour. Calcined flint is used as a decorative aggregate for white concrete cladding panels.

**122** Mortar containing beach sand fine aggregate (Mustique, Saint Vincent, and the Grenadines) composed of feldspars (grey/white) and pyroxenes (brightly coloured); XPT, ×35.

**123** Polymictic marine sand fine aggregate (Oman) with eight constituents. Consisting chiefly of serpentinized peridotite (brightly coloured) and shell, with minor peridotite, pyroxene, quartz, chert, sandstone, and calcrete; XPT, ×35.

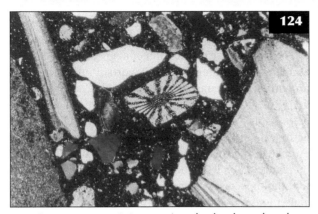

**124** Concrete containing marine dredged sand and gravel aggregate. Note the presence of marine shell fragments (light brown). A sea urchin spine is seen in cross-section in the centre of view; XPT, ×35.

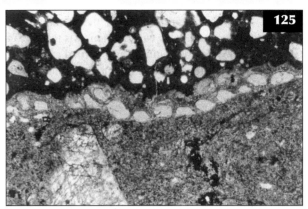

**125** Concrete containing marine dredged coarse aggregate. A coarse aggregate particle (lower) exhibits encrustation by a marine organism (centre); PPT, ×35.

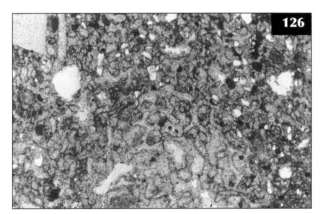

**126** Coral aggregate (Tanzania) composed of calcareous material formed as coralline skeletal structure; PPT, ×35.

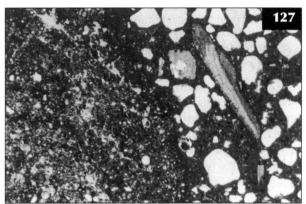

**127** Concrete containing 'Lytag' lightweight coarse aggregate (left). This consists of roughly spherical particles with a porous interior and an outer sealed surface; PPT, ×35.

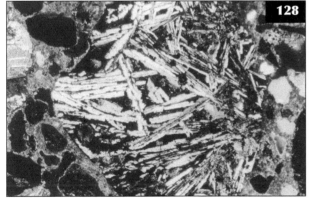

**128** Concrete containing boiler clinker aggregate (centre). The clinker is porous with an opaque matrix and acicular crystals of melilite (white); XPT, ×35.

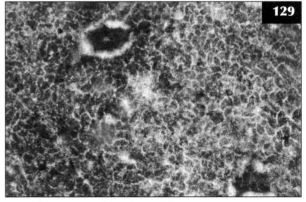

**129** Close view of a calcined flint coarse aggregate particle, showing 'tile structure' characteristic of the mineral cristobalite; PPT, ×150.

Recycled aggregates are being used increasingly in the current drive for sustainable construction. It is possible to use demolition waste as engineering fill, road aggregate, or concrete aggregate (Building Research Establishment, 1998). Recycled aggregates typically comprise crushed concrete and/or brick masonry. Special care is required to ensure that they do not also contain potentially deleterious contaminants such as glass, bituminous material, timber, gypsum, metal, plastic, and clay lumps. The approximate composition of recycled aggregate can be determined by hand separation. Detailed investigations to determine the nature of the constituents should be conducted using petrographic examination. One of the key quality parameters in assessing the likely performance of recycled concrete aggregate is the quantity of cement mortar (cement and fine aggregate) that remains on the surface of the coarse aggregate. Abbas *et al.* (2007) have developed a petrographic/image analysis method to measure this.

## AGGREGATE GRADING, SHAPE, AND SURFACE TEXTURE

Aggregates are divided into coarse aggregate and fine aggregate on the basis of particle size. The European standard (British Standards Institution, 2002a) defines coarse aggregate being >4 mm size and fine aggregate <4 mm size. The American standard (ASTM International, 2008b) defines coarse aggregate being >4.75 mm size

and fine aggregate <4.75 mm size. An 'all-in' aggregate consists of a mixture of coarse and fine aggregates.

The particle size distribution of aggregate (also known as aggregate 'grading') has an important influence on its properties as a construction material. The grading is definitively classified by sieve testing but the petrographer will need to estimate the grading as part of petrographic examination. Aggregate grading may be estimated by visual comparison with the chart provided in Figure 130. It should be noted that the standard terminology used for aggregate grading in the construction industry differs from geological classification of sediments, with a well graded aggregate corresponding to moderately sorted sediment. Figure 131 shows a well graded fine aggregate, while Figure 132 shows a uniformly graded fine aggregate.

Aggregate particles within a particular size fraction may exhibit a variety of different shapes, which affects the workability and particle interlock of aggregate. Surface textures of aggregate particles are important in determining workability and adhesion to binders. Aggregate particle shape and surface texture may be classified using the chart in Figure 133. In addition to the terms used in Figure 133, other terms appropriate for use when describing surface texture include glassy, smooth, granular, rough, crystalline, honeycombed, and porous.

Good concrete aggregate has rounded particles, to slide over each other and make concrete more workable

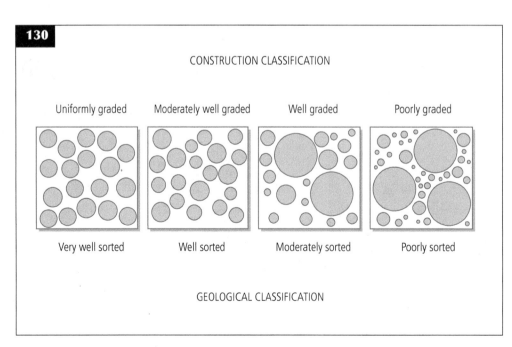

**130** Comparison chart for classification of aggregate particle grading by visual assessment (Ingham, 2005).

as it is poured, and a rough surface texture, to improve the bonding with the cement matrix. A well graded aggregate needs less cement, because the smaller particles fill the spaces between the larger ones. Flaky particles make for a harsher mix of lower workability at a given water content than nonflaky ones. This leads to poor compaction and a high void content resulting in low strength and durability. Rocks that give rise to excessively elongate or flaky aggregate particles include gneiss, schist, amphibolite, phyllite, and slate. The optimum properties for concrete aggregate are found in natural sand and gravel resources. Crushed rock, being angular, is less suitable, but is used in areas where gravels are scarce.

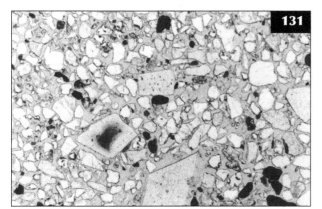

**131** Well graded fine aggregate or 'sharp sand'. The aggregate is also polymictic consisting chiefly of quartz with minor flint (brown) and ironstone (black); PPT, ×35.

**132** A uniformly graded fine aggregate or 'soft sand'. The aggregate is also monomictic consisting wholly of quartz grains; XPT, ×35.

**133** Comparison chart for classification of aggregate particle shape and surface texture (Ingham, 2005).

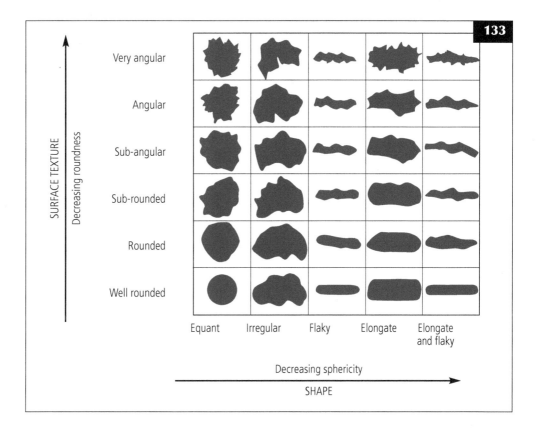

Fine aggregate for use in Portland cement mortars, renders, and screeds should have rounded particles ('soft sand' or 'building sand') to improve the workability of the mix (see 132). Fine aggregates with angular particles ('sharp sand') are suitable for the fine aggregate fraction of concrete or road aggregate (see 131).

'Silver sand' is sand with a very low iron content that is favoured for the manufacture of high-quality optical glass. Its light colour ('silver') is a result of consisting almost entirely of quartz and there being very little contamination with other minerals (134). It has limited uses in the construction industry.

Road aggregate is used to make up the various layers of road pavements (see Chapter 10). The sub-base is a drainage layer with coarse, uniformly graded aggregate particles with large interparticle spaces, to prevent capillary rise of water. The roadbase bears and spreads the traffic load and ideally needs strong, rounded, continuously graded particles (<40 mm maximum sized) that give good compaction. The upper two layers of the pavement, the basecourse and wearing course, are always bitumen bound. This means that for the aggregate (<20 mm maximum sized), a rough surface texture and angular shape to the particles are important as well as strength. Crushed igneous rocks and coarse-grained, indurated sandstones make the best wearing course aggregates.

## SOUNDNESS, IMPURITIES, AND UNDESIRABLE CONSTITUENTS OF AGGREGATES

Natural aggregates exhibit considerable variability. For example, a single granite quarry can contain compositional variants, textural variants, hydrothermally altered zones, weathered zones, mineralized zones, dykes, metamorphic enclaves, sheared lithologies, cataclasite, and tuffisite (French, 1994). This has important consequences for the potential occurrence of deleterious materials and for the investigation of these materials.

The soundness of a rock may be decreased through natural weathering (135) or alteration (136). Geological weathering typically results in an increase in porosity and water absorption. Depending on the degree of weathering,

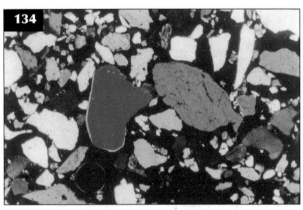

**134** 'Silver sand' consisting almost entirely of fine quartz grains. As the sand was derived from the weathering of granite there are traces of contamination with tourmaline (orange); XPT, ×35.

**135** Slightly weathered granite (weathering grade II) exhibiting slight discolouration and slight weakening; PPT, ×35.

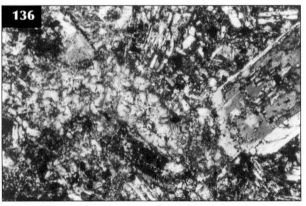

**136** Hydrothermally altered andesite (St Vincent, Caribbean) showing chloritization of ferromagnesian minerals. Chlorite appears green and blue; XPT, ×150.

this can affect the response of the rock in service to its environment should such rock be used to make aggregates. Degree of weathering may be assessed petrographically using the classification shown in *Table 14*.

Other undesirable aggregate constituents include:

*Porous aggregate* – these increase the water demand of the mix when used in concrete, which has a detrimental effect on concrete strength and durability. Highly microporous rocks may be frost susceptible. Flints often have a highly microporous white cortex and may cause 'pop-outs' on concrete surfaces when exposed to freeze–thaw processes (**137**). Chalk is distinctive limestone that outcrops in western Europe (with chalk-like rock found in other locations) and it occurs as a constituent in sand and gravel deposits. Chalk is typically weak, highly microporous, and is considered to be an unsuitable aggregate constituent in all but trace amounts (**138**).

**Table 14** Classification of weathered rock for engineering purposes (from Anon, 1995; a similar table appears in BS 5930:1999a)

| Grade | Classifier | Typical characteristics |
|---|---|---|
| I | Fresh | • Unchanged from the original state |
| II | Slightly weathered | • Slight discolouration, slight weakening |
| III | Moderately weathered | • Considerably weakened, penetrative discolouration<br>• Large pieces cannot be broken by hand |
| IV | Highly weathered | • Large pieces can be broken by hand<br>• Does not readily disaggregate (slake) when a dry sample is immersed in water |
| V | Completely weathered | • Considerably weakened<br>• Slakes<br>• Original texture is apparent |
| VI | Residual soil | • Soil derived by *in situ* weathering but retaining none of the original texture or fabric |

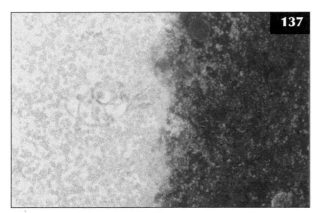

**137** Flint aggregate particle showing the frost-susceptible highly microporous cortex (left, yellow) and the nonporous core (right, dark green); UV, ×150.

**138** Concrete fine aggregate (southeast England) contaminated with chalk (brown); XPT, ×35.

*Clay, silt, and dust* – fines are defined as any materials passing the 63 µm sieve in Europe (BS EN 12620) or the 75 µm sieve in America (ASTM C33). Excessive fines in aggregates (**139**) can have a number of detrimental effects. In concrete, fines will increase the water demand of the mix and consequently ASTM C33 limits the fines content of coarse aggregate for concrete to 1% and fine aggregate to 3% (concrete subject to abrasion), 5% (concrete not subject to abrasion, or crushed rock subject to abrasion), or 7% (crushed rock, concrete not subject to abrasion). Dust coatings on aggregate particles inhibit adhesion of binders such as bitumen or cement. It is important to ascertain if fines contain any deleterious swelling clays. This is probably best undertaken using XRD analysis.

*Mica* – micas (muscovite, biotite, chlorite) are commonly found in certain rocks used for aggregate (e.g. granite, gneiss) and discrete mica flakes may be released when aggregate is crushed (see **310**). Free mica is undesirable in concrete/mortar aggregate, as it will increase the water demand. Compressive strength of Portland cement mortar has been found to be lowered by 20% with a 2% mica content and 40% with a 4% mica content (Leemann & Holzer, 2001).

*Chlorides* – these occur naturally in marine dredged (and certain coastal and land sourced) aggregates, usually as sodium chloride. Chlorides can initiate reinforcement corrosion in concrete. Although chloride ions cannot be seen through the microscope, evidence of a marine origin of aggregate (such as the presence of marine shells) can alert the petrographer to a potential for chloride contamination. The chloride content should then be determined by chemical analysis.

*Shell* – calcareous shell fragments are found in sand and gravel aggregates from marine and coastal deposits (see **284**). Due to their shape, shell fragments increase the water demand of concrete aggregate. BS 882 (British Standards Institution, 1992) limits the shell content of aggregate for concrete to 20% (by mass) for <10 mm fractions, and 8% for fractions of >10 mm.

*Organic matter* – these are potentially deleterious impurities for concrete aggregates as certain types can retard/prevent the setting of cement. Figure **140** shows a sandstone with a bituminous matrix that was proposed (and rejected) for use as coarse aggregate. Figure **141**

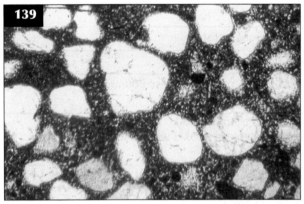

**139** Natural fine aggregate consisting of quartz grains and a high proportion of fines (dark brown); PPT, ×150.

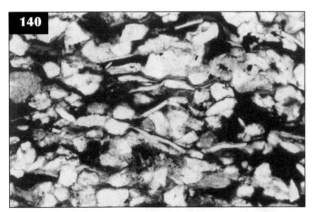

**140** Bituminous sandstone consisting of quartz grains (white) and muscovite mica flakes (white, elongated) with a matrix that includes abundant bitumen (black/brown); PPT, ×150.

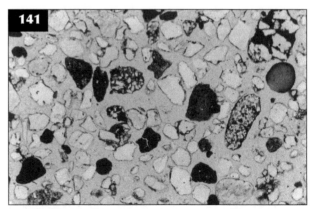

**141** Natural fine aggregate for concrete that comprises quartz grains (white) and organic matter (brown); PPT, ×35.

shows a fine aggregate that is contaminated with organic material. Other forms of organic matter, such as coal and lignite (see **217**), are undesirable in concrete aggregate as they are weak, unsound, and can cause staining at concrete surfaces. ASTM C33 limits the amount of coal and lignite in concrete aggregate to 0.5% for concrete elements with visible surfaces and 1% for hidden concrete.

*Alkalis* – high levels of alkali metal ions (sodium and potassium) can cause alkali–aggregate reaction (AAR) in concrete (in the presence of reactive aggregate and water). Salt-contaminated aggregates (marine aggregate) can contain alkalis that can potentially contribute to ASR. A number of common rock-forming minerals (feldspars, phyllosilicates) contain appreciable alkalis and in certain rare cases these have been released in concrete, causing ASR (Goguel & Milestone, 2007). AAR is discussed further in the following section and on pp.109–114.

*Sulfides and sulfates* – iron pyrite (iron sulfide) occurs in some natural aggregate deposits such as the flint gravels of southeast England. Some pyrites oxidize when incorporated in concrete aggregate and may cause 'pop-outs' and brown staining on concrete surfaces (see **216**). Reactive pyritic aggregate particles can be distinguished from unreactive ones by immersing them in a limewater solution. Reactive pyrite particles will form a blue–green gelatinous precipitate within 5 minutes that changes to a brown colour within 30 minutes (Midgley, 1958). Sulfate minerals are present in some aggregate resources and when incorporated into concrete may cause sulfate attack. In areas of the Middle East (French & Crammond, 1980) and in tropical coastal plains, sulfates commonly occur in aggregate resources (**142**).

## POTENTIAL ALKALI-REACTIVITY OF AGGREGATE FOR CONCRETE

AAR is the expansive reaction between alkali (sodium and potassium) hydroxides in the pore solution of concrete and minerals in the aggregates. These are deleterious reactions that cause expansive cracking of concrete structures, which detrimentally affects their durability.

Petrographic examination is the ideal starting point for the assessment of aggregates for alkali–aggregate reactivity (Sims *et al.*, 2002). In the United Kingdom, concrete aggregate combinations are classified as being of 'low', 'normal', or 'high' reactivity on the basis of their petrographic composition. An aggregate or aggregate combination is classified as being of low reactivity if it comprises 97% or more of rock and mineral constituents considered to be of low reactivity potential. An aggregate or aggregate combination is classified as being of high reactivity if it comprises either more than 10% crushed greywacke or recycled demolition waste. An aggregate is classified as being of normal reactivity if it cannot be classified as being of low or high reactivity. Any aggregate containing detectable opal or opaline silica should not be classified and, in most circumstances, should not be used. A list of aggregate and mineral types

**142** Limestone coarse aggregate particle (Saudi Arabia) including the sulfate mineral gypsum (grey). The thin section has been stained in accordance with Dickson's method and the nonferroan calcite of the limestone appears pink; XPT, ×35.

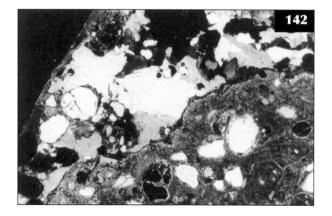

and their respective alkali-reactivity classification is given in *Table 15*.

In ASR, there are two classes of minerals that are known to be highly reactive in concrete. These are metastable and disordered forms of quartz such as opal, chalcedony (143), tridymite (see 355), and cristobalite (129), and alumino-silicate glasses in the matrix of intermediate to acid volcanic rocks. These minerals may be present in rocks that would otherwise be classified as low or normal reactivity and care must be taken to ensure that these are detected if present. Crushed greywacke, a type of poorly sorted sandstone with >15% fine-grained matrix, has also proved to be highly alkali–silica reactive when used as concrete aggregate (Huenger & Weidmueller, 2007) (144, and see 219).

Limestones are usually of low alkali–silica reactivity but care must be taken with silicified limestones as these may be reactive due to the presence of chert (145).

As mentioned previously, natural aggregates exhibit considerable variability with a wide range of different lithologies potentially being present. As a result, an alkali-reactive minor constituent may be present where the main lithology is regarded as highly satisfactory and inert. Certain lithologies may contain disordered or submicroscopic forms of quartz that are reactive. Taking the example of our granite quarry from p. 66, the typical granite would be of low reactivity (146), but a microgranite textural variant may contain alkali-reactive disordered silica (147). Different shear zones within the quarry have dynamically metamorphosed the granite

**Table 15** Alkali–silica reactivity of rock, mineral, and artificial concrete aggregate constituents (compiled from EN 7943 and BRE Digest 330)

| Low reactivity | Normal reactivity | High reactivity |
|---|---|---|
| Air-cooled blastfurnace slag | Arkose [4] | Chalcedony or chalcedonic silica |
| Amphibolite [1] | Breccia | Cristobalite |
| Andesite [1] | Chert | Greywacke (crushed) |
| Basalt [1] | Conglomerate | Opal or opaline silica |
| Chalk | Flint | Tridymite |
| Diorite and microdiorite [1] | Granulite [1] | Recycled demolition waste |
| Dolerite [1] | Greywacke (uncrushed) | |
| Dolomite [2] | Gritstone [4] | |
| Expanded clay/shale/slate | Hornfels [1] | |
| Feldspar | Quartzite [4] | |
| Gabbro [1] | Rhyolite [1] | |
| Gneiss [1] | Sandstone and siltstone [4] | |
| Granite and microgranite [1] | Tuff [1] | |
| Limestone [2] | Volcanic glass [5] | |
| Marble [2] | | |
| Quartz [3] | | |
| Schist [1] | | |
| Sintered pulverized-fuel ash | | |
| Slate | | |
| Syenite and microsyenite [1] | | |
| Trachyte [1] | | |

1   Rock type generally not alkali–silica reactive but may occasionally contain reactive forms of silica (possibly including opal, chalcedony, tridymite, devitrified glass, microcrystalline or cyptocrystalline quartz, submicroscopic or disordered quartz).
2   Silicified limestones, dolomites, and marbles may be reactive due to the presence of disseminated microcrystalline or cyptocrystalline quartz.
3   Not quartzite, nor microcrystalline or cryptocrystalline quartz. In addition, highly disordered quartz and submicroscopic silica or is potentially reactive.
4   Sandstones may occasionally contain reactive forms of silica such as microcrystalline or cyptocrystalline quartz, strained or disordered quartz, opal, chalcedony. Greywacke can be highly alkali reactive (when crushed) and is considered separately from other sandstones.
5   Volcanic glass may be reactive if the noncrystalline glass has devitrified to become very finely crystalline.

**143** Chalcedony (fibrous silica) within a flint aggregate particle; XPT, ×150.

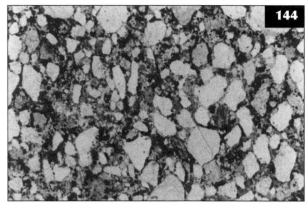

**144** Greywacke coarse aggregate particle. Quartz grains appear white and the matrix appears brown; PPT, ×35.

**145** Silicified limestone with microcrystalline silica appearing grey and calcite shown pink/brown; XPT, ×300.

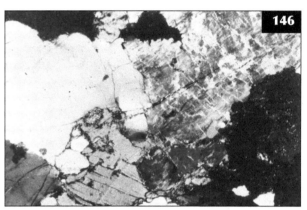

**146** Granite coarse aggregate particle consisting of coarsely crystalline quartz (grey, left), feldspar (grey, right), and biotite mica (brown); XPT, ×35.

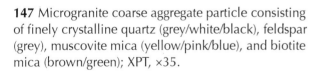

**147** Microgranite coarse aggregate particle consisting of finely crystalline quartz (grey/white/black), feldspar (grey), muscovite mica (yellow/pink/blue), and biotite mica (brown/green); XPT, ×35.

variously to cataclasite (**148**), ultracataclasite (**149**), and ultramylonite (**150**). These rocks contain higher proportions of disordered quartz and are likely to be alkali-reactive (Wigum, 1996).

Quartzite and chert (and flint) are important rock types when considering aggregate for concrete, as they are both commonly used in concrete and are potentially alkali-reactive. Classification of these rocks can be challenging and with this in mind, Hunt (1995) has devised an engineering classification of quartzite and chert to aid alkali–silica reactivity assessment. The classification is based on texture, fabric, and grain size definitions as shown in *Table 16*. Using this classification, Figure 151 shows diaquartzite, Figure 152 shows metaquartzite, Figure 153 shows cherty quartzite, and Figure **77** shows chert.

AAR is discussed further on pp. 109–114.

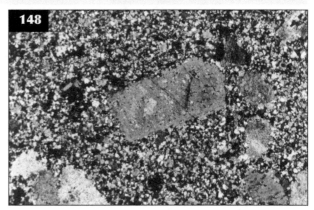

**148** Granite cataclasite with a porphyritic texture with porphyroclasts of feldspar (centre) set in a groundmass largely composed of finely crystalline quartz; XPT, ×35.

**149** Ultracataclasite consisting of an unfoliated groundmass of finely crystalline to cryptocrystalline silica, with rare strained porphyroclasts; XPT, ×35.

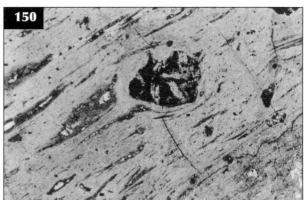

**150** Ultramylonite with a foliated groundmass of finely crystalline to cryptocrystalline silica, with rare strained porphyroclasts; XPT, ×35.

**Table 16** Classification of the quartzite group of rocks (from Hunt, 1995)

| FABRIC | TEXTURE | | | Content of nanocrystalline quartz (<2 µm) |
|---|---|---|---|---|
| | (Ortho)quartzite | Diaquartzite | Metaquartzite | 0% |
| | Cherty quartzite | Cherty diaquartz | Cherty metaquartz | 5% |
| | Quartzitic chert | Quartzitic diachert | Quartzitic metachert | 50% |
| | Chert | Diachert | Metachert | 95% / 100% |
| | Original grain structure visible | No original grain structure, interlocking crystals | Interlocking crystals, visible restructuring | |

———————— INCREASING MATURITY ————————→

**151** Diaquartzite consisting of interlocking crystals of medium-grained quartz; XPT, ×150.

**152** Metaquartzite consisting of quartz that has been visibly sheared, restructured, and includes submicroscopic quartz; XPT, ×150.

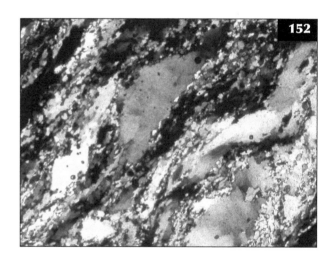

**153** Cherty quartzite consisting of a mixture of cryptocrystalline quartz (chert, right) and interlocking medium-grained quartz crystals (quartzite, left); XPT, ×150.

## MATCHING AGGREGATES FOR CONSERVATION

When conserving and restoring historic buildings it is often necessary to find a match for aggregate used originally in historic mortars, plasters, and renders. Data from petrographic examination can be used to pinpoint the original source by comparing the petrographic data with geological and historical information. The original source may no longer be operational and it may be impossible/inappropriate to commence new extractive works due to planning restrictions. In these circumstances a good match may be found by comparison with that of commercially available material. This is a visual process, supplemented by petrographic examination findings, with the main aggregate characteristics considered being colour (often controlled by mineralogy), grading, and particle shape.

To aid the process of locating aggregate sources, lists of suppliers and test data for some currently available building aggregates suitable for use in restoration have been published. These include the English Heritage Directory of Building Sands and Aggregates (2000) that lists approximately 60% of the sources in England. Information regarding Scottish aggregate sources can be found in Scottish Aggregates for Building Conservation (Leslie & Gibbons, 1999). Addresses of other potential sources in the United Kingdom can be found in the Directory of Mines and Quarries (British Geological Survey, 2002) or the Directory of Quarries and Quarry Equipment (Quarry Management, 2007/2008).

# Concrete

## INTRODUCTION

Concrete is a mixture of aggregate, water, and a binder (today usually Portland cement-based). The binder is chemically activated by the water to form a paste, which sets the inert aggregate into a hardened stone-like mass. Today, concrete is the most commonly used construction material with present global consumption estimated at 11 billion tonnes annually (Metha & Monteiro, 2006). It owes its popularity to being relatively cheap and readily available, that it can be poured into a variety of element shapes and sizes, and that it is water resistant. Concrete elements are manufactured either '*in situ*' using prefabricated formwork or are 'precast' in moulds.

The use of concrete-like materials dates back thousands of years, with early concrete binders including nonhydraulic lime, hydraulic lime, and lime mixed with pozzolana. However, the invention of the binder most commonly used today is credited to John Aspin, who in 1824 patented a cement produced by firing a mix of ground limestone and clay that would set under water. Aspin considered his cement to resemble the Jurassic Portland stone of Dorset (England) and called his product 'Portland cement'. This is the predominant cement variety in use today.

Plain concrete is inherently strong in compression but relatively weak in tension. For this reason, steel reinforcement bars are often cast into the concrete to improve the tensile strength (reinforced concrete). With prestressed concrete, by tensioning steel tendons, a precompression is introduced to counteract tensile stresses and prevent cracking. Modern concretes also contain a number of chemical admixtures and/or mineral additions to impart desirable properties to the concrete, either in the fresh state or post-hardening. In Europe, concrete is specified in accordance with EN 206-1 (British Standards Institution, 2000b) and in America it is specified to ASTM C94/C94M (ASTM International, 2009).

The principal applications of petrographic examination to concrete investigations are:
- Identifying the mix ingredients including coarse and fine aggregate, cement type, fillers, mineral additions, and certain additives.
- Determining the mix proportions and air void content.
- Assessing the quality of workmanship.
- Screening for evidence of distress or deterioration.
- Diagnosing the causes, severity, and extent of defects and deterioration.

## ASSESSMENT OF CONCRETE STRUCTURES

Concrete structures require assessment for a variety of reasons, which will usually be associated with investigation of either specification compliance, maintenance requirements, or structural adequacy. It is essential to establish the precise aims of inspection and testing before planning the assessment programme. The typical elements of an assessment programme include documentation review, preliminary site visit, access and safety provision, selection of test methods/numbers/ locations, visual inspection, testing, interpretation, and documentation of findings. Useful information regarding assessment and testing procedures is provided by Kay (1992), Concrete Society Technical Report No. 54 (Concrete Society, 2000) and Bungey *et al.* (2006).

Petrographic examination has an important role in the assessment of concrete structures and the petrographer should be involved in planning investigations. The petrographer should be fully aware of the construction details, environmental conditions, and history of the structure under investigation. Ideally, they would have the opportunity to visit the site and be involved in the visual inspection and selection of sample locations. Failing this, all relevant documentation should be made available including drawings, photographs, inspection records, and test data.

# PETROGRAPHIC EXAMINATION AND COMPLEMENTARY TECHNIQUES

Petrographic examination of concrete is performed in accordance with ASTM C856 (ASTM International, 2004). At the time of writing a European standard was being developed but was not yet available. Useful reference handbooks for concrete petrography have been written by St John *et al.* (1998) and Walker *et al.* (2006).

Following arrival in the laboratory, core or lump samples are first examined in the as-received condition, both with the unaided eye and using a low-power binocular microscope at magnifications of up to ×100. Finely ground slices of concrete may be prepared to aid the visual/low-power microscopical examination and/or for micrometric determination of the mix proportions by point counting. This initial examination is used to observe macroscopic features such as coarse aggregate type, cement matrix colour, relative hardness, mixing, compaction, and macrocracking. It also allows selection of the most appropriate location for thin sections to be taken for further, more detailed high-power microscopical examination. Owing to the presence of coarse aggregate, the minimum size of thin sections should be 75 mm × 50 mm and larger sizes such as 100 mm × 75 mm are desirable. For most investigations, two thin sections are sufficient, one from the outer surface and one from the interior at depth. It should be noted that the outer 50 mm of concrete elements are not considered to be representative.

Thin sections are examined in plane-polarized or cross-polarized transmitted light using a medium- to high-power petrological microscope at magnifications typically up to ×600. This is used to determine the mineralogy of the aggregate, the cement type, the presence of mineral additions/pigments, assess the quality of workmanship, and screen the material for evidence of distress or deterioration. Fluorescent dye is usually added to the consolidating resin during sample preparation to aid the examination of cement matrix microporosity and cracks/microcracks, when the specimen is viewed in ultraviolet light (fluorescence microscopy). To determine if sulfate-resisting Portland cement is present, it is necessary to prepare a highly polished and etched (with hydrofluoric acid [HF] vapour or potassium hydroxide [KOH] solution) specimen for examination under reflected light.

A range of complementary methods is available to the concrete petrographer. The most commonly used are SEM and XRD. SEM requires highly polished specimens to be prepared or alternatively a thin section without a glass cover slip. SEM allows closer microscopical examination and the on-board EPM system provides inorganic bulk elemental chemical analysis either of selected points, or an area up to 30 mm × 15 mm ('elemental mapping'). XRD requires a powdered sample and, for crystalline minerals, actually identifies the minerals present, rather than just giving an elemental composition. Thermal analysis methods such as differential thermal analysis (DTA), thermogravimetric analysis (TGA) and differential scanning calorimetry (DSC) are occasionally used for analysing cement and concrete. Infrared spectroscopy is used for identifying and quantifying organic materials present in concrete.

# INVESTIGATING THE COMPOSITION AND QUALITY OF CONCRETE

## OVERVIEW

Good quality of concrete mix design, ingredients used, and workmanship are fundamental to achieving strong and durable concrete structures. Petrographic examination is routinely used to determine concrete composition, either to check compliance with a specification or to provide baseline information for asset engineering. By petrographically examining a concrete sample, it is possible to observe directly the geological type and characteristics of coarse and fine aggregate, the cement type used, and the presence of mineral additions, fillers, and fibres. The presence of concrete ingredients that cannot be directly observed may be suggested by observable properties that they impart to the concrete (e.g. air-entraining chemical admixtures). Micrometric analysis of petrographic specimens can be used to determine the cement content, air void content, and water/cement ratio (W/C) of hardened concrete. Workmanship issues that can be assessed include the adequacy of mixing, effectiveness of compaction, and curing (Ingham & Hamm, 2005).

## AGGREGATES TYPES AND PROPERTIES

Aggregate makes up the major part of concrete (typically 75% by volume) and the aggregate properties have a considerable affect on its engineering and aesthetic characteristics. In Europe, concrete aggregates are specified in accordance with EN 12620 (British Standards Institution, 2002), while in America, ASTM C33/C33M (ASTM International, 2008a) is used. Most concrete mix designs specify an aggregate that is continuously graded (well graded) in order to minimize the interparticle space to be occupied by the cement. An aggregate from a single source with a suitable grading may be used as an 'all-in' aggregate. However, it is more common blend a coarse aggregate with a fine aggregate; the dividing line between coarse and fine being set at 4 mm particle size for CEN standards and 4.75 mm for ASTM standards.

The petrographer should concentrate their aggregate description on those features that affect concrete properties. Unless the aggregate is 'all-in', the coarse and fine aggregate fractions are described separately. Features described should include aggregate type, nominal maximum size, shape, grading, distribution, and orientation. Each rock type present should be identified and the colour, relative hardness, percentage of total, and degree of alteration or weathering should be described. The relative hardness of aggregate constituents (and cement matrix) can be determined empirically in hand specimen using the following classification, which is based on Mohs' Scale of Hardness:

- Very soft – can be penetrated easily by a finger.
- Soft – scores with a fingernail.
- Moderately soft – scores using a copper coin.
- Moderately hard – scores easily with a penknife.
- Hard – not easily scored with a penknife.
- Very hard – cannot be scored with a steel point or knife.

Visual estimation of the relative proportions of different aggregate constituents using a comparitive chart such as in Figure 19, is sufficient for most purposes. If required, more accurate quantifications may be achieved by point-counting, either in thin section or of finely ground slices.

The petrographic identity of aggregate indicates its source and the petrographic description will allow detection of the source region, or possibly even the specific source quarry. It is important to remember that the coarse and fine aggregates may be derived from different sources.

The aggregate type may be natural gravel, marine dredged, crushed rock, artificial, or recycled. Natural gravels are indicated by rounded or subrounded shape, although some angular/partly angular particles will be present where oversized gravel particles have been crushed down during processing. Figures 154 and 155 show concretes containing natural sand and gravel aggregates from two different river terrace deposits in England. Figure 156 shows a concrete from the Middle East that contains a natural wadi gravel coarse aggregate that has been partially crushed. Marine dredged gravel

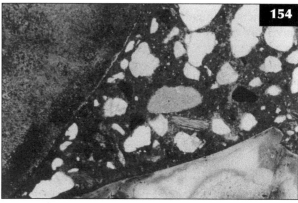

**154** Concrete with natural sand and gravel aggregate from the Thames Valley river terraces (England), comprising rounded flint coarse aggregate particles (brown, lower and left) and quartz (white), ironstone (black), and limestone (light grey) fine aggregate particles; PPT, ×35.

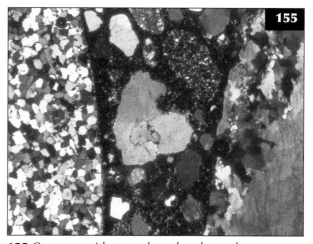

**155** Concrete with natural sand and gravel aggregate from the River Trent terraces (England), comprising rounded quartzite coarse aggregate particles (left and right) and quartzitic fine aggregate particles (grey/white); XPT, ×35.

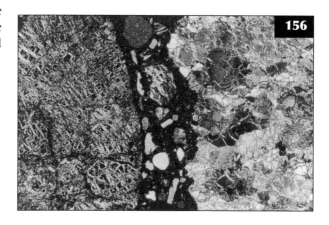

**156** Concrete (from the United Arab Emirates) containing partially crushed wadi gravel coarse aggregate. Rounded serpentinized peridotite (grey/left) and angular processed peridotite (right) coarse aggregate particles are seen; XPT, ×35.

aggregate can often be distinguished from terrestrial gravels by the presence of unfossilized calcareous shell fragments and encrusting organisms (see 124 and 125). The use of crushed rock aggregate is indicated by particles with angular shape and often by monomictic composition. Figure 157 shows concrete with a crushed gabbro coarse aggregate and Figure 158 shows concrete containing crushed limestone coarse aggregate.

Some artificial aggregates used in concrete may have beneficial characteristics, while others are undesirable. Figure 159 shows lightweight concrete made with sintered PFA coarse aggregate, which is a low density aggregate. Figure 160 shows white concrete from a

cladding panel, incorporating flint coarse aggregate that has been calcined to give it a decorative white colour. Figure 161 shows concrete where boiler clinker waste has been used as aggregate. It is important to identify the presence of boiler clinker as it may contain high levels of deleterious salts that can corrode metal elements of the construction, such as steel reinforcement or steel beams.

The aggregate grading (particle size distribution) is an important characteristic as it affects the workability and packing density of the concrete mix. The coarse aggregate grading can be assessed petrographically from core specimens and finely ground slices, while the fine aggregate grading is best assessed in thin section. The

**157** Concrete with crushed rock (gabbro) coarse aggregate and quartz fine aggregate particles. The uncarbonated cement matrix appears black; XPT, ×35.

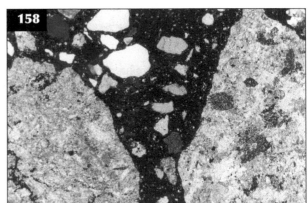

**158** Concrete with crushed limestone coarse aggregate (light brown, left and right) and a quartz-rich (grey/white) natural sand fine aggregate. The uncarbonated cement matrix appears black; XPT, ×35.

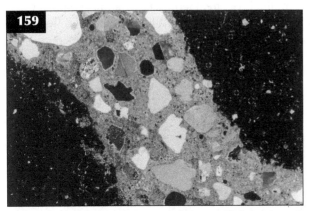

**159** Lightweight concrete containing artificial sintered PFA coarse aggregate (black/isotropic) and natural quartz sand fine aggregate (grey/white). The cement matrix is fully carbonated (light brown); XPT, ×35.

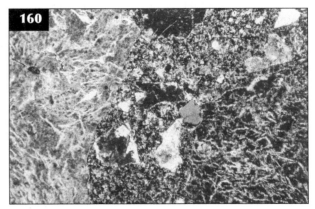

**160** Decorative concrete containing crushed calcined flint aggregate (brown) and traces of natural sand, including quartz (white) and glauconite (green); PPT, ×35.

main types of grading for coarse aggregates are continuous, single-size, and gap-graded. Fine aggregate gradings can be described using the chart in Figure 130. Aggregate shape and surface texture affect concrete mix workability and adhesion of the cement paste respectively. Aggregate particle shape and surface texture may be classified using a standard chart such as in Figure 133. Additional terms for describing surface texture include glassy, smooth, granular, rough, crystalline, honeycombed, and porous.

The petrographer should identify the presence and quantity of undesirable or deleterious aggregate constituents including porous particles, flaky particles, shell, excessive fines, mica, organic matter, sulfate and sulfide-bearing particles, and potentially alkali-reactive constituents. The effects of many of these constituents will be discussed in subsequent sections and further information regarding aggregates is provided in Chapter 4.

## PORTLAND-TYPE CEMENTS

Portland cement is the predominant type of binder used in concrete. Portland cement is manufactured by heating limestone and aluminosilicate rock (such as clay or shale) in a rotary kiln at temperatures of up to 1500°C. This produces cement 'clinker' which is interground with around 5% gypsum (which acts as an early set retarder) to give the cement product.

Various types of Portland cement can be produced depending on the composition of the raw materials, their relative proportions, and the heating/cooling regime. The fineness of the cement can also be varied to provide different rates of hardening and strength gain. Both 'pure' and 'composite' varieties of Portland cement are available. The main 'pure' varieties are normal Portland, sulfate-resisting, and white Portland cement. In 'composite' cements a proportion of the Portland cement clinker is replaced by industrial byproducts such as a blastfurnace slag, fly ash, natural pozzolana, or limestone.

In Europe, Portland cements are specified in accordance with EN 197-1 (British Standards Institution, 2000a) and a summary of this classification is shown in *Table 17* (overleaf). In America, Portland cements are specified using ASTM C150 (ASTM International, 2007) and composite (blended) cements using ASTM C595 (ASTM International, 2008b), as summarized in *Table 18* (overleaf).

In terms of composition, Portland cement clinker is dominated by four main phases, these being tricalcium silicate ($C_3S$), dicalcium silicate ($C_2S$), tricalcium aluminate ($C_3A$), and tetracalcium aluminoferrite ($C_4AF$). Their microscopical properties are summarized in *Table 19* (overleaf). These clinker phases are hydraulic and on mixing with water they react to form a range of hydrated phases called calcium silicate hydrate gels (C-S-H), which are largely responsible for the binding action of the cement in concrete. Accounts of the hydration reactions

**161** Filler joist floor concrete containing boiler clinker aggregate (black) including vesicular and glassy forms and some residual furnace fuel (black, left). The cement matrix is shown brown; PPT, ×35.

**Table 17** Summary of European EN 197-1 classification of Portland cement types

| Cement type | Notation | | Constituents | |
|---|---|---|---|---|
| | | | Cement clinker, % | Addition, % |
| CEM I | Portland cement | CEM I | 95–100 | – |
| CEM II | Portland–slag cement | CEM II/A-S | 80–94 | 6–20 blastfurnace slag |
| | | CEM II/B-S | 65–79 | 21–35 blastfurnace slag |
| | Portland–silica fume cement | CEM II/A-D | 90–94 | 6–10 silica fume[1] |
| | Portland–pozzolana cement | CEM II/A-P | 80–94 | 6–20 natural pozzolana |
| | | CEM II/B-P | 65–79 | 21–35 natural pozzolana |
| | | CEM II/A-Q | 80–94 | 6–20 natural calcined pozzolana |
| | | CEM II/B-Q | 65–79 | 21–35 natural calcined pozzolana |
| | Portland–fly ash cement | CEM II/A-V | 80–94 | 6–20 siliceous fly ash |
| | | CEM II/B-V | 65–79 | 21–35 siliceous fly ash |
| | | CEM II/A-W | 80–94 | 6–20 calcareous fly ash |
| | | CEM II/B-W | 65–79 | 21–35 calcareous fly ash |
| | Portland–burnt shale cement | CEM II/A-T | 80–94 | 6–20 burnt shale |
| | | CEM II/B-T | 65–79 | 21–35 burnt shale |
| | Portland–limestone cement | CEM II/A-L | 80–94 | 6–20 limestone |
| | | CEM II/B-L | 65–79 | 21–35 limestone |
| | | CEM II/A-LL | 80–94 | 6–20 limestone |
| | | CEM II/B-LL | 65–79 | 21–35 limestone |
| | Portland–composite cement | CEM II/A-M | 80–94 | 6–20 of two more of any mineral addition materials in this table |
| | | CEM II/B-M | 65–79 | 21–35 of two more of any mineral addition materials in this table |
| CEM III | Blastfurnace cement | CEM III/A | 35–64 | 36–65 blastfurnace slag |
| | | CEM III/B | 20–34 | 66–80 blastfurnace slag |
| | | CEM III/C | 5–19 | 81–95 blastfurnace slag |
| CEM IV | Pozzolanic cement | CEM IV/A | 65–89 | 11–35 of two more of silica fume[1], pozzolana, or fly ash |
| | | CEM IV/B | 45–64 | 36–55 of two more of silica fume[1], pozzolana, or fly ash |
| CEM V | Composite cement | CEM V/A | 40–64 | 18–30 blastfurnace slag and 18–30 of either pozzolana or fly ash |
| | | CEM V/B | 20–38 | 31–50 blastfurnace slag and 31–50 of either pozzolana or fly ash |

1   The proportion of silica fume is limited to 10%.

**Table 18** Summary of American classification of Portland cement types

| ASTM C150 cement type | Cement description |
|---|---|
| Type I | Normal Portland cement |
| Type II | Modified cement |
| Type III | Rapid hardening Portland cement |
| Type IV | Low heat Portland cement |
| Type V | Sulfate-resisting cement |
| **ASTM C595 cement type** | **Blended cement description** |
| Type IS | Portland blastfurnace slag cement |
| Type IP | Portland–pozzolan cement |
| Type P | Portland–pozzolan cement for use when higher strengths at early ages are not required |
| Type I(PM) | Pozzolan-modified Portland cement |
| Type I(SM) | Slag-modified Portland cement |
| Type S | Slag cement |
| | |

**Table 19** Properties of the principal components of cement clinker

| Property | Component | | | |
|---|---|---|---|---|
| Mineral | Tricalcium silicate (alite) | Dicalcium silicate (belite) | Tricalcium aluminate (aluminate) | Tetracalcium aluminoferrite (brownmillerite or ferrite) |
| Abbreviated cement chemists' formula | $C_3S$ | $C_2S$ | $C_3A$ | $C_4AF$ |
| Full chemical formula | $3CaO.SiO_2$ | $2CaO.SiO_2$ | $3CaO.Al_2O_3$ | $4CaO.Al_2O_3.Fe_2O_3$ |
| Typical proportion in cement (wt %) | 55–65 | 15–25 | 8–14 | 8–12 |
| Relief | High | High | High | Very high |
| Colour in PPT light | Colourless to slightly coloured | Colourless, yellow, or green | Brown in Portland cement and colourless in white cement | Brown to yellow. Pleochroic, green to almost opaque |
| Birefringence | First-order grey interference colours | Second-order interference colours | None to low to yellow | First-order white |
| Form | Lath, tablet-like or equant. Six sided in cross-section. 25–65 μm in size | Rounded grains, often clustered | Usually cubic form filling spaces between belite and ferrite crystals | Bladed, prismatic, dendritic, fibrous, massive or infilling |
| Extinction and twinning | Wavy to straight, length slow extinction. Rare polysynthetic twinning | Lamellar twinning common | May have oblique, length slow extinction. Twinning may be present | Crystals exhibit length slow extinction |
| Colour in reflected light after etching with KOH solution | Does not etch | Does not etch | Bluish-grey | White |
| Colour in reflected light after etching with HF vapour | Buff brown | Blue | Light grey | White |

of cement are given in Bye (1999), Taylor (1997), and Barnes and Bensted (2002). When studying cement reactions it is important to know that a shorthand system called 'cement chemists' notation' is used describe cement compounds. This uses single letters to abbreviate the usual oxide formulae as shown in *Table 20*.

The hydrated cement paste within concrete contains residual grains of unhydrated/partially hydrated clinker and these can be examined microscopically to determine the cement type. In modern cements, residual cement grains are chiefly small (<20 µm) with sporadic medium-sized (20–60 µm) and rare large grains (60–100 µm) sometimes present. In older concrete (pre-1950s), Portland cement was more coarsely ground and very large relict cement grains (>100 µm) are commonly observed. Figures **162–164** show the appearance in thin section of a large residual unhydrated cement grain in ordinary Portland cement concrete. The cement grain has an appearance reminiscent of a bunch of grapes, consisting of agglomerated phenocrysts of calcium

**Table 20** Cement chemists' notation.

| Oxide | CaO | SiO$_2$ | Al$_2$O$_3$ | Fe$_2$O$_3$ | H$_2$O | Na$_2$O | K$_2$O | SO$_3$ | MgO | CO$_2$ |
|---|---|---|---|---|---|---|---|---|---|---|
| Symbol | C | S | A | F | H | N | K | $\bar{S}$ | M | $\bar{C}$ |

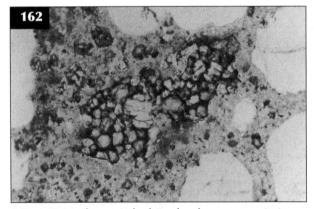

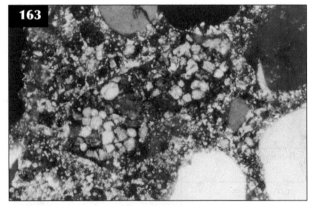

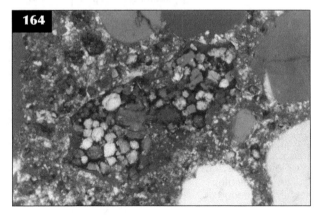

**162–164** Modern residual Portland cement grain in thin section specimen viewed in PPT (**162**), XPT (**163**), and in XPT with the gypsum plate inserted into the light path (**164**). The cement grain is surrounded by uncarbonated cement matrix and quartz fine aggregate particles (white in PPT); ×150.

silicates ($C_3S$ and $C_2S$) with the other clinker constituents ($C_3A$ and $C_4AF$) forming the fine-grained interstitial component. In plane-polarized light, the calcium silicates are largely colourless with the $C_3S$ (alite) consisting of euhedral pseudohexagonal crystals, while the $C_2S$ (belite) forms anhedral or subhedral crystals that sometimes have a rounded appearance. The interstitial phases are typically brown in colour due to the iron content of the $C_4AF$ (brownmillerite). This iron content gives normal Portland cement its characteristic light grey colour. Figures **165** and **166** show the appearance in thin section of a very large residual cement grain in a pre-1950s concrete. In plane-polarized light, the $C_2S$ (belite) crystals have a distinct green colour, while the $C_3S$ (alite) crystals are colourless and the interstitial phases are brown.

To resist sulfate attack, buried concrete may incorporate sulfate-resisting Portland cement (SRPC), which has a low $C_3A$ (aluminate) content. This reduces the deleterious reaction between $C_3A$ and sulfates from soil/groundwater that forms calcium sulfoaluminate (ettringite). The British standard for SRPC (BS 4027; 1996) stipulates a maximum $C_3A$ content of 3.5% and otherwise SRPC is similar to normal Portland cement. In America, SRPC is known as 'type V' cement and ASTM C150 limits $C_3A$ content to 5%. By reducing the $C_3A$ content of SRPC during its manufacture the $C_4AF$ (brownmillerite) content increases correspondingly. In hand specimen, this gives hardened SRPC a slightly darker shade of grey than normal Portland cement. Figure **167** shows concrete made with SRPC, in thin section. In PPT, residual SRPC grains apparently exhibit

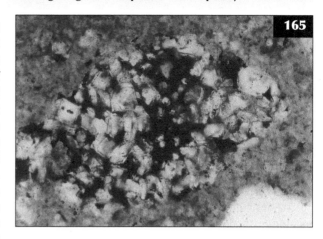

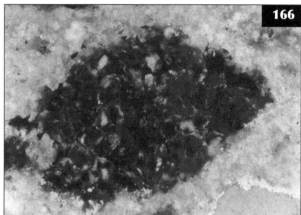

**165, 166** Historic residual Portland cement grain in thin section specimen viewed in PPT (**165**) and XPT (**166**). The cement grain is surrounded by carbonated cement matrix (light brown in XPT). A quartz fine aggregate particle is seen lower right; ×300.

**167** Residual grain of sulfate-resisting Portland cement (centre) seen in thin section. Calcium silicates ($C_2S$ and $C_3S$) appear colourless and the interstitial phases (chiefly $C_4AF$) appear dark brown; PPT, ×300.

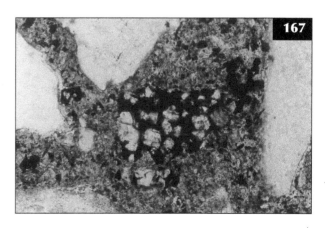

a higher proportion of the brown-coloured $C_4AF$ interstitial phase than normal Portland cement. However, thin section examination alone is not a reliable indicator of the use of SRPC, it must be confirmed by reflected light examination of a highly polished and etched specimen. The method is described in BS 1881: Part 124 (British Standard Institution, 1988) and by Grove (1968). For reflected light examination of cements, the choice of etchant reflects the constituent to be identified and a range of appropriate etching techniques are described in Appendix B. For differentiating between normal and sulfate-resisting Portland cements the two most useful etching methods are those that use either HF vapour or 10% KOH solution. These two different etching techniques produce the following colours that are seen in reflected light:

| Clinker phase | HF vapour | 10% KOH |
|---|---|---|
| $C_3S$ (alite) | Brown | Does not etch |
| $C_2S$ (belite) | Blue | Does not etch |
| $C_3A$ (aluminate) | Light grey | Bluish grey |
| $C_4AF$ (ferrite) | White | White |

The method for distinguishing between SRPC and normal Portland cement is based on the ratio of $C_3A$ to $C_4AF$ in the interstitial phases of residual cement clinker grains. For normal Portland cement the ratio does not usually exceed 1:2 (aluminate:ferrite) and in contrast the ratio is typically more than 1:5 for SRPC. A least 20 residual cement grains >20 μm in size should be examined for each concrete or cement sample. Figures **168–170** show residual cement grains within concrete polished specimens that have all been etched with HF vapour for 2–5 seconds. Figure **168**

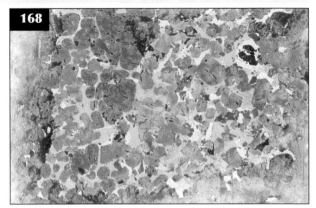

**168** Unhydrated Portland cement clinker after etching with HF vapour. $C_3S$ is brown, $C_2S$ is blue (very little in this example), the interstitial matrix is differentiated into $C_3A$ (light grey) and $C_4AF$ (white); PPR, ×300. (Courtesy of Barry J Hunt.)

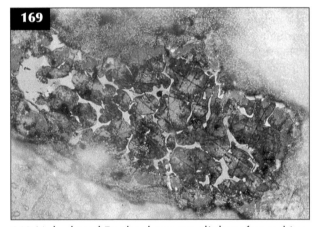

**169** Unhydrated Portland cement clinker after etching with HF vapour. $C_2S$ is blue and exhibits lamellar twinning, $C_3S$ is brown. The interstitial matrix is differentiated into $C_3A$ (light grey) and $C_4AF$ (white); PPR, ×300. (Courtesy of Barry J Hunt.)

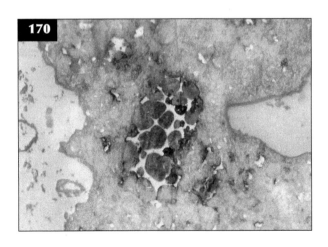

**170** Unhydrated sulfate-resisting Portland cement clinker after etching with HF vapour (centre). $C_3S$ is brown, $C_2S$ is blue (very few in this example), the interstitial phases consist almost entirely of $C_4AF$ (white); PPR, ×300. (Courtesy of Barry J Hunt.)

shows a normal Portland cement grain with the calcium silicates mainly consisting of $C_3S$ (alite). Figure **169** shows a normal Portland cement grain with the calcium silicates mainly consisting of $C_2S$ (belite). Figure **170** shows an SRPC grain with the interstitial phases predominantly consisting of $C_4AF$ (ferrite). A comprehensive illustrated reference for the examination of cement clinker is provided by Campbell (1999).

It is important to note that owing to its low $C_3A$ content, SRPC is more susceptible to chloride ion ingress as $C_3A$ binds with chloride ions to form calcium chloroaluminate, making them unavailable to initiate reinforcement corrosion. This increased chloride ingress of SRPC is a major disadvantage for reinforced concrete that is exposed to chlorides, such as in coastal or highways structures.

White or coloured concrete is sometimes required for architectural purposes. White Portland cement is used to provide white or light-coloured concrete, in combination with suitably coloured aggregates. White Portland cement is manufactured from very pure white limestone and china clay, which have a very low iron content (<0.3%) and may be burned at temperatures in excess of 1600°C (Bye, 1999). In thin section, white Portland cement is readily identified as the unhydrated clinker has a near absence (<1%) of ferrite ($C_4AF$) and a high content of alite ($C_3S$). Figure **171** shows a concrete made with white Portland cement with the lack of brown ferrite in the interstitial phases being apparent. The content of aluminate ($C_3A$) is similar or slightly higher than in normal Portland cement clinker. White Portland cement is more expensive, has a lower alkali content, and typically lower strength than normal (grey) Portland cement. Coloured concrete is produced using coloured cement that usually comprises white Portland cement with an added pigment. Red, yellow, brown, or black cements can be produced by intergrinding 5–10% (by weight) iron oxide pigment of the corresponding colour. Green cements can be made by using chromium oxide and blue-coloured cements are made using cobalt blue pigment. An example of a mortar incorporating yellow-coloured cement is shown in Figure **318**.

In thin section (**172**), the hardened Portland cement matrix of uncarbonated concrete is seen to consist mainly (approximately 70%) of essentially amorphous

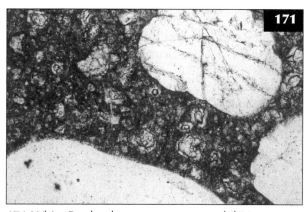

**171** White Portland cement concrete exhibiting a frequent abundance of unhydrated and partially hydrated (with hydration rims visible) residual white Portland cement grains (white) in the cement matrix (brown). Quartz fine aggregate particles appear white/rounded; PPT, ×150.

**172** Close view of Portland cement concrete showing the cement matrix (black) which contains brightly coloured portlandite crystallites (white/yellow) and relict cement grains (dark brown). Flint (upper/grey) and quartz (grey/white) fine aggregate particles are also visible; XPT, ×150.

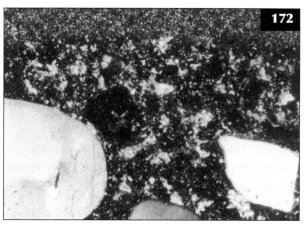

and isotropic hydrated calcium silicate gel (C-S-H), which is also referred to as 'tobermornite gel'. Within this are embedded crystallites of portlandite (calcium hydroxide, approximately 10%) and residual unhydrated cement grains. In thin section, portlandite $(Ca(OH)_2)$ is visible in the cement matrix as anhedral or euhedral crystallites, plates or short prisms, up to 100 μm in size. They are colourless in plane-polarized light and in cross-polarized light they are highly birefringent and stand out against almost isotropic uncarbonated gel. Over time (years to decades) the cement matrix gradually reacts with the atmosphere causing carbonation which alters the matrix to crystalline calcium carbonate (see p. 96).

When describing the hardened cement matrix of concrete samples the petrographer would normally include the following. In hand specimen describe the colour, distribution, relative hardness, and any evidence of distress or deterioration Microscopically, the features described should include the cement type, the nature of relict cement grains, and the nature of portlandite crystallites. Description of the relict cement grains and portlandite crystallites should include size, relative abundance, distribution, and shape. Size and relative abundance can be described using the following classifications:

- Size (average measurement across the grains or crystallites):
  - Small – <20 μm.
  - Medium – 20–60 μm.
  - Large – 60–100 μm.
  - Very large – >100 μm.
- Relative abundance:
  - Rare – only found by thorough searching.
  - Sporadic – only occasionally observed during normal examination.
  - Common – easily observed during normal examination.
  - Frequent – easily observed with minimal examination.
  - Abundant – immediately apparent to initial examination.

## ADDITIONS AND ADMIXTURES

Additions are inert, pozzolanic, or latent hydraulic mineral powders, which are added at the mix. Admixtures are chemicals added to concrete at the time of mixing. These should not be confused with additives, which are chemicals preblended with the cement or dry cementitious mix. Other materials added to concrete include fibres which are most commonly composed of steel or plastic.

It is now common practice for a proportion of the cement in concrete to be replaced by one or more mineral additions. When combined with mineral additions, cements are referred to as 'composite' or 'blended' cements, in Europe and the USA respectively. Cost savings gained from replacing cement clinker with industrial waste products was the original reason for the development of blended cements. Now mineral additions are used to confer a range of performance benefits to the concrete mix and to reduce the carbon footprint of concrete binders.

Mineral additions include a range of natural and industrially manufactured materials that often exhibit pozzolanic behaviour. The original pozzolana was volcanic pumice from Pozzouli, Italy (**173**) that was used by the Romans in lime concrete/mortar (see also p. 148). The Romans discovered that when added to lime, the pozzolana would allow it to set underwater, thus enabling marine construction. This was due to the pozzolanic behaviour, which involves a reaction between amorphous silica in the pozzolana and calcium hydroxide in the lime binder or cement clinker to form C-S-H. For certain pozzolanas, the pozzolanic reaction is slower than Portland cement hydration, with correspondingly lower rates of heat liberation and strength gain. For Portland cement concrete, this has the advantage of reducing the risk of early thermal cracking but with the disadvantage of increasing the curing time. The binder resulting from use of mineral additions is often more durable as the pozzolanic reaction confers reduced permeability (to aggressive agents) and a lower content of calcium hydroxide (that could otherwise have participated in deleterious reactions).

In modern concrete, the most commonly used

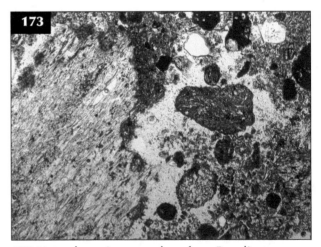

**173** Natural pumice pozzolana from Bacoli, near Pozzuoli (Italy); PPT, ×75.

pozzolanic mineral additions are pulverized fuel ash (PFA), ground granulated blastfurnace slag (GGBS), condensed silica fume (CSF or microsilica), and metakaolin. *Table 21* shows the typical addition rates and advantages and disadvantages of these four main types. Fly ashes (including PFA) are byproducts of coal burning power stations. They largely comprise spherical, 1–150 μm diameter particles of isotropic aluminosilicate glass, with some iron-bearing phases and traces of carbonaceous unburnt fuel. In concrete thin sections, fly ash additions are readily detected by the presence of unreacted glassy spheres and cenospheres (hollow spheres) that may vary in colour from yellow, brown, red, black, or grey, and traces of opaque unburnt coal (**174**). In hand specimen, concrete containing a PFA addition may have a darker coloured cement matrix (medium grey instead of light grey).

**Table 21** Generalized proporties of mineral additions for concrete in comparison to concrete made with Portland cement alone (data mainly from Bamforth, 2004)

| Property | Pulverised fuel ash (PFA) | Ground granulated blastfurnace slag (GGBS) | Silica fume (CSF) | Metakaolin |
|---|---|---|---|---|
| Typical addition rate by % of cement | 20–40 | 40–70 | 5–10 | 8–15 |
| Workability | Reduced water demand | Water demand may be slightly reduced | Increases water demand, plasticising admixture required | Increases water demand, plasticicing admixture required |
| Heat of hydration | Reduced | Reduced | Increased | Increased |
| Rate of strength gain | Rate reduced | Rate reduced with higher final strength | Rate increased with much higher final strength | Rate increased with higher final strength |
| Carbonation | Similar | Increases in some situations | Similar | Similar |
| Resistance to chloride ingress | 3–10 times better | 3–10 times better | Improved | Marked Improvement |
| Sulfate resistance | Improved | Improved | Improved resistance to sodium sulfate but not magnesium sulfate | Improved |
| ASR prevention | Improved | Improved | Improved except where opaline aggregate present | Improved |
| Frost resistance | May be improved at higher addition rates | May be more susceptible to surface scaling | Insufficient data | Insufficient data |

Blastfurnace slag (including GGBS) is a byproduct of the iron-making process that is rapidly cooled (during either granulation or pelletization) to form slag with a glassy disordered structure. In hand specimen, concrete containing a GGBS addition exhibits a distinctive dark green cement matrix, which with time changes to a cream colour as the sulfides in the GGBS oxidize. In thin section, the GGBS is readily detected by the presence of isotropic glassy shards of unreacted GGBS that are typically 1–100 μm in size (**175**). The shards are usually transparent and colourless, or light brown/green (or sometimes white or dark brown) and sporadic, opaque grains or inclusions of iron may be present.

It may be necessary to determine the GGBS or PFA content of a blended cement product. If the individual ingredients are available separately this can be achieved by point-counting of highly polished specimens in reflected light. A number of reference specimens should be made up with a range of known GGBS/PFA contents and these should be point-counted along with the blended cement sample in order to estimate the relative accuracy of the determination. It will be necessary to etch the specimens (e.g. with 1% nitric acid solution) to distinguish the mineral addition from the cement clinker. Microscopically determining the quantity of GGBS or PFA in hardened concrete is much more difficult as much of the mineral addition will be too small to resolve optically or will have been consumed by pozzolanic or hydration reactions. The most reliable techniques involve using scanning electron microscopy and EPM of highly polished specimens, as described by French (1991b, 1992).

Silica fume (CSF or microsilica) is a byproduct of the electric arc furnace manufacture of silicon or ferrosilicon alloys from high purity quartz and coal. The process gives off fumes consisting of extremely small spheres (<0.1– 1 μm diameter) of very pure amorphous silica that is a very reactive pozzolana. The extreme fineness presents handling difficulties so it is usually supplied in the form of pellets or slurry. Detecting silica fume additions by microscopy is difficult owing to its extreme fineness, low addition rate (typically 5–10%), and the fact that almost all particles are consumed by the pozzolanic reaction. However, its presence may be disclosed by agglomerations of silica fume if dispersion during concrete mixing has been incomplete (**176**). In addition, the cement matrix may appear optically dense and portlandite depleted. Metakaolin is another highly reactive pozzolanic mineral addition. It is manufactured from deposits of naturally occurring kaolin clay by heating in a kiln at 700–800°C. As with silica fume it is unlikely to be directly detected by microscopy, for the same reasons.

Ternary (triple blend) cements comprise Portland

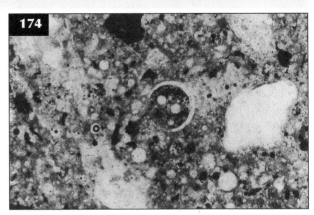

**174** Close view of concrete containing PFA. The cement matrix exhibits unreacted glassy spheres (white, brown) and cenospheres (centre) of PFA and traces of unburnt coal (black); PPT, ×300.

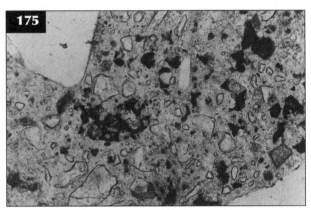

**175** Close view of concrete containing GGBS. The cement matrix exhibits angular, glassy 'shards' of GGBS (transparent) and traces of iron (black); PPT, ×300.

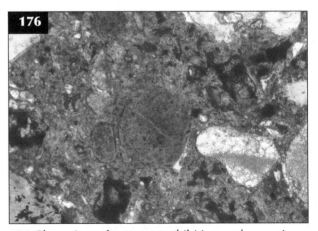

**176** Close view of concrete exhibiting agglomerations of incompletely dispersed silica fume (light brown, centre); XPT, ×200. (Courtesy of Mike Eden.)

cement with two mineral additions to produce high-strength or high-performance concrete. These are usually blends of cement and silica fume (typically around 5%) with either GGBS or PFA as the third component (20–50%). While the petrographer will easily observe the presence of GGBS or PFA, the silica fume will most likely go undetected in thin section.

Inert limestone fillers are being increasingly used in concrete (added to the cement or at the mixer). EN 197-1 (British Standards Institution, 2000a) contains two types of Portland–limestone cement containing either 6–20% or 21–35% limestone filler. In thin section, limestone filler is observed as highly birefringent calcium carbonate dust (typically <75 μm sized particles). When the concrete contains limestone aggregate, care must be taken not to confuse limestone filler with dust of fracture from the aggregate (**177**).

Modern concrete mixes are usually designed to include one or more chemical admixtures to modify the fresh or hardened concrete properties. Admixtures are supplied as powders or aqueous solutions and dosage levels are low, typically being 0.3–1.5% by weight of cement. Admixtures are used variously to improve workability and lower the water demand of the mix (plasticizers/superplasticizers), retard or accelerate the rate of setting, confer air entrainment, or add waterproofing qualities.

Chemical admixtures cannot be directly observed using optical microscopy, although their presence may sometimes be inferred from their effect on the concrete.

For example, air-entraining admixtures will confer the presence of substantial quantities of small (10 μm to 1 mm diameter), spherical air voids, throughout the cement matrix (see **185**). Infrared spectroscopy is the method of choice for identifying the presence of organic chemical admixtures in concrete. However, given the relatively small dosage rates of admixtures, their detection cannot be guaranteed. Determination of the quantity of admixture used is particularly difficult and requires comparison with use of specially formulated reference samples.

Fibre-reinforced concrete contains discontinuous discrete fibres to reduce shrinkage and improve a variety of properties, including flexural toughness, impact resistance, and resistance to spalling and delamination. Fibres of various shapes and sizes made from either steel, plastic, glass, or natural materials are used. The most common are steel macrofibres which usually comprise <2% of the concrete. Figure **178** shows fibre-reinforced concrete from a floor slab that incorporates round steel macrofibres (0.75 mm diameter). The appearance of polypropylene microfibres is illustrated in Figure **327**.

## WATER/CEMENT RATIO

The ratio of water to cement in a concrete mix has a considerable influence on the strength and durability of the hardened concrete. For a fully compacted concrete, its strength is inversely proportional to the water/cement ratio (W/C) (Neville, 1995). Water is added to the mix, firstly, to hydrate the cement and, secondly, to make the concrete

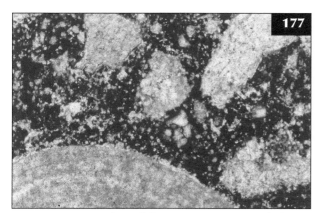

**177** Concrete made with limestone aggregate (light brown). The uncarbonated cement matrix appears black and contains limestone dust from the aggregate (light brown) and relict cement grains (grey/yellow); PPT, ×150.

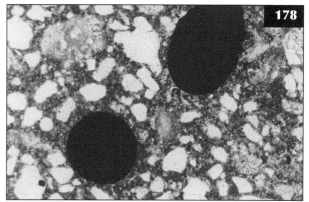

**178** Fibre-reinforced concrete containing round steel macrofibres (black). One fibre (lower left) is seen in cross-section while another (upper right) is seen in oblique section; PPT, ×35.

workable enough to compact. The water of hydration chemically binds in the cement hydrates, while the water for workability occupies a system of capillary pores (around 1 μm in size) within the concrete (with some being lost by bleeding and evaporation). As a general rule, the higher the W/C, the more microporous and consequently the less durable, the hardened concrete will be.

The W/C of hardened concrete is usually determined either as part of routine condition assessment for concrete structures, or as an aid to determining the cause of unsatisfactory concrete performance. It is of particular importance where poor concrete performance is suspected or alleged to have been caused by incorrect or improper W/C, for example where too high a W/C has resulted in inadequate compressive strength. Although there is no internationally recognized test, original W/C is most commonly determined using indirect chemical analysis in accordance with methods such as that described in BS 1881: Part 124 (British Standards Institution, 1988). The original W/C is obtained by combining determined values for the water bound in the cement hydrates and the capillary porosity. The chemical method is unsuitable for air-entrained, cracked, carbonated, or poorly compacted concrete and even in favourable circumstances (W/C in the range 0.4–0.8) it only has an accuracy and reproducibility of within 0.1 of the actual W/C (Concrete Society, 1989).

The relative inaccuracy of the chemical methods has led to the development of alternative petrographic methods. These are direct methods based on petrographic indicators of apparent W/C. In general as W/C increases, the hardened cement paste exhibits a higher microporosity, a lower concentration of residual unhydrated cement grains, and more (and larger) portlandite crystallites. For normal Portland cement concrete, an experienced petrographer should be able to use these microscopical features to estimate visually apparent W/C to one of the following broard ranges:

| Apparent W/C | Range |
| --- | --- |
| <0.35 | Low |
| 0.35–0.65 | Normal |
| >0.65 | High |

This approximate determination of W/C is often sufficient as the petrographer is usually only required to establish whether or not a particular concrete had a W/C that significantly exceeded the target value (St John *et al.*, 1998).

If more accurate W/C determination is required, for example in the case of a dispute, it is necessary to compare the thin section with reference specimens

prepared from concretes with a range of W/C (e.g. 0.30, 0.35, 0.40, 0.45, 0.50, 0.55, 0.60, 0.65, 0.70, 0.75, 0.80), but with similar ingredients to the sample under investigation. With reference specimens available there are two methods used for W/C determination. The most commonly used involves observing the relative fluorescence using fluorescent microscopy and the other method involves observing the relative amounts of residual unhydrated cement grains.

Apparent W/C determination by fluorescence microscopy is widely used and a standard method is available, which is described in Nordisk NT 361-1999. The sample is first impregnated using fluorescent yellow dye and a thin section of the sample is prepared to a thickness of 20–25 μm (rather than the standard 30 μm). The thin section is subsequently examined microscopically in fluorescent light (either transmitted or reflected). The intensity of the fluorescence emitted from the cement paste is proportional to the amount of intruded resin, which is in turn related to the capillary porosity and the W/C. Figures **179–181** illustrate the typical relative fluorescence emitted by hardened concrete samples with W/C in the low, normal, and high ranges respectively. The cement paste may appear homogeneous or show large variation in microporosity. Large variations can result from improper mixing or bleeding. Figure **182** shows concrete that exhibits a large variation in microporosity of the cement paste making determination of the apparent W/C challenging. In Denmark, an image analysis software package (W/C-check) has been developed for the determination of apparent W/C from camera images of fluorescent impregnated thin sections when viewed under the microscope (**183**).

The accuracy of the fluorescence microscopy method of determining W/C is believed to be at least as good as, and usually better than, the chemical analysis method. Jakobsen *et al.*, (2003) claim an accuracy of ±0.02 but this is likely to be achievable in only the most favourable circumstances with excellent reference samples. For unknown concrete from existing structures the accuracy is more likely to be within 0.1 (Neville, 2003). It is not yet clear whether the relative fluorescence of the dyed resin used for impregnation of W/C reference specimens will degrade with time as the thin sections are stored. Sibbick *et al.* (2007) report that they have found no change in their reference specimens after 15 years.

Another direct method for determining W/C described by French (1991a) involves the determination of the number of residual clinker particles per millimetre traverse of paste. The sample is compared with reference

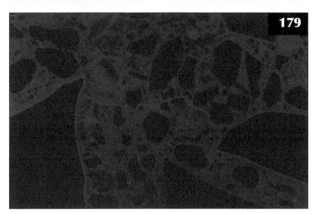

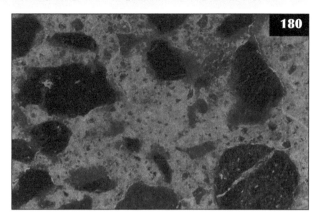

**179–181** Concrete samples with W/Cs in the low (<0.35), normal (0.35–0.65), and high (>0.65) ranges for **179**, **180**, and **181** respectively. Using fluorescence microscopy the microporosity of the cement paste appears low (**179**, dark green), normal (**180**, light green), and high (**181**, yellow). Fine aggregate particles appear black; UV, ×150.

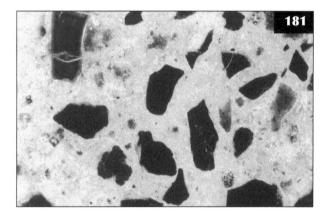

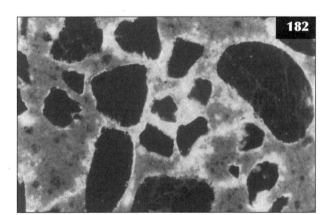

**182** Concrete with variable microporosity of the cement matrix ranging from normal (light green) to high (yellow). Overall W/C is difficult to determine but is probably in the normal range (0.35–0.65); UV, ×150.

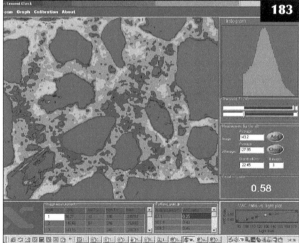

**183** A user interface of W/C-CHECK software for automatic determination of W/C of concrete thin section specimens, showing a colour segmented image, intensity histogram, calibration curve, and analysis data. (Courtesy of Concrete Experts International.)

thin sections visually and can be quantified by measuring the number of unhydrated clinker grains encountered on a traverse of at least 120 mm in length. French suggests that the results are accurate to ±0.02. A similar method, described by Ravenscroft (1982) involves determination of the unhydrated clinker by point-counting. This method is more time consuming and uses highly polished and etched specimens of cement paste that are examined in reflected light.

It is useful to know that a number of different methods using scanning electron microscopy and associated microanalysis have been developed for determining W/C. These are yet to be widely used and variously involve:

- Comparison of chemical (oxide) composition of the hardened cement matrix determined using the EDS (energy dispersive X-ray spectroscopy) attached to the electron microscope with standard specimens (French, 1991a).
- Examination of the grey level observed in backscattered imaging mode. The grey level appears darker as W/C increases (Gonçalves *et al.*, 2007).
- Quantifying the volumetric fractions of capillary pores, hydration products, and unreacted cement using high resolution FESEM in the backscattered electron mode. The free W/C is calculated from the obtained data and the volumetric expansion coefficient of cement hydration (Wong & Buenfeld, 2007).

**184** Surface of a finely ground slice of concrete (100 × 75 mm area) that is obliquely illuminated to pick out the air voids (in shadow). Most of the air voids in view are entrapped air voids and the compaction is good (excess voidage in the 0.5–3.0% range). Crushed limestone aggregate appears light brown while the cement matrix is shown medium brown.

### AIR VOIDS

When freshly mixed concrete is placed, it is typically compacted (by vibration) to eliminate entrapped air bubbles and improve aggregate packing. However, some air bubbles will invariably be retained to become entrapped air voids in the hardened concrete. In addition, air-entraining chemical admixtures may be added to the fresh concrete to improve frost resistance of the hardened concrete, by facilitating the formation of entrained air voids. Also, the use of water-reducing chemical admixtures (superplasticizers) may result in a lesser degree of air entrainment.

Entrapped air voids (**184**) are mostly >1 mm across and are typically irregular in shape. They are irregularly distributed in concrete and often increase in number and size towards the concrete surface. In contrast, entrained air voids (**185**) are between 10 μm and 1 mm in diameter, are characteristically spherical in shape, and are uniformly distributed throughout the concrete. Deliberately entrained air may form 4–5% of the volume of concrete.

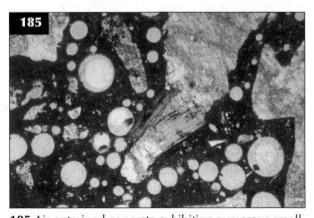

**185** Air-entrained concrete exhibiting numerous small, spherical entrained air voids (yellow). Crushed granite aggregate appears mottled and the cement matrix is shown dark brown; PPT, ×35.

The degree of concrete compaction is normally assessed by estimating the excess voidage. The excess voidage is the volume percentage of entrapped air voids present over and above that found in the same concrete which has been as fully compacted as possible. This is normally estimated visually in hand specimen by comparison with standard specimens, or more commonly standard photographs, such as those provided in Concrete Society Technical Report No. 11 (1987). Once the excess voidage has been estimated the degree of compaction can be described thus:

| Excess voidage, % | Compaction |
|---|---|
| >0.5 | Very good |
| >0.5–3.0 | Good (normal for satisfactory quality structural concrete) |
| >3.0–5.0 | Medium |
| >5.0–10.0 | Poor |
| >10.0 | Very poor |

In addition, the term 'honeycombed' is used to describe interconnecting large entrapped air voids (excess voidage may be as high as 30% locally) arising from inadequate concrete compaction or a lack of mortar (binder and fine aggregate).

For accurate determination of the air void system, it is necessary to undertake manual point-counting or linear traverse measurements of finely ground concrete slice specimens. In Europe, determination of air void content of hardened concrete is performed in accordance with EN 480: Part 11 (British Standards Institution, 2005) and in America, ASTM C457 (ASTM International, 2008c) is used. It is important to ensure that the sample is adequately representative of the concrete. The minimum area of concrete slice for concrete with a 20 mm coarse aggregate would be 10,000 mm² (and greater for concrete with larger aggregate). The finely ground slice specimen is carefully prepared, ensuring a scratch-free surface with well defined air void edges. The volume proportions of aggregate, cement matrix, entrapped air voids, and entrained air voids are determined by examining the slice using a low-power binocular microscope with the aid of an oblique incident light source and a mechanical stage. The air void content is then calculated using formulae supplied in the EN or ASTM standards.

The manual method of air void analysis takes several hours to perform for each specimen. With this in mind, an automated apparatus, the RapidAir 457 has been developed to perform ASTM C457 or EN 480-11 air void analysis in less than 15 minutes (Jakobsen *et al.*, 2005). The apparatus comprises a computerized control unit (and monitor) with image analysis software, a video camera, and a microscope objective mounted on a moving stage (**186**). The concrete sample preparation for image analysis is similar to that required for the manual method, except that the contrast between the air voids and the concrete has to be enhanced by colouring (**187**).

## MODAL ANALYSIS OF CONCRETE BY PETROGRAPHIC PROCEDURES

Modal analysis of hardened concrete (also known as 'micrometric' analysis) may be used to determine binder content, aggregate content, and aggregate grading. As with air void determination, this is performed on finely ground concrete slice specimens (see **184**) that must be large enough to represent adequately the concrete element from which it is taken (typically 100 × 100 mm minimum area). The normal methods involve either linear traverse or point-counting determinations to establish the volume proportions of coarse aggregate, fine aggregate, cement paste, and air voids. As with

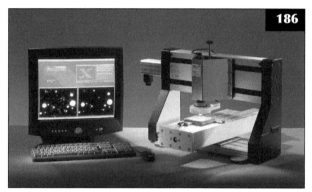

**186** The RapidAir 457, Automated Air Void Analyzer system. (Courtesy of Concrete Experts International.)

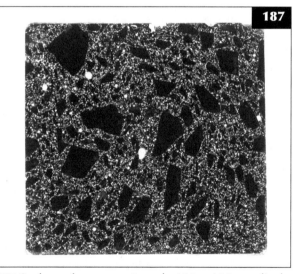

**187** Surface of an approximately 100 × 100 mm finely ground slice of air-entrained concrete after contrast enhancement, to allow air void analysis using the RapidAir 457. Entrapped and entrained air voids appear white against the black background of aggregate and cement matrix. (Courtesy of Concrete Experts International.)

point-counting for air void analysis, alternative image analysis techniques are being developed in an effort to make modal analysis more efficient. From the volume data, the original mix proportions (by weight) can be calculated in a number of different ways. One approach is detailed in Concrete Society TR No. 32 (1989) and reproduced in St John *et al.* (1998).

Modal analysis is a suitable alternative to chemical analysis methods for determining cement content and aggregate grading, such BS 1881: Part 124 (British Standards Institution, 1988). Modal analysis may be preferable in circumstances where chemical analysis may be subject to inaccuracies from the presence of acid-soluble aggregate constituents (limestone etc.) or chemically altered concrete (leaching and other deleterious reactions). Experience to date suggests that binder content determinations by modal analysis are of comparable accuracy (or better) to chemical analysis (Concrete Society, 1989).

## WORKMANSHIP

General concrete workmanship issues that can be assessed by petrographic examination include compliance with specified mix proportions, the adequacy of mixing, and effectiveness of placing, compaction, curing, and finishing. A variety of specific workmanship issues may need to be investigated as the result of a perceived inadequacy of the finished concrete product. For example, insufficient strength and durability resulting from extra water being added to the mix at the site can be checked for by assessing the W/C, using fluorescence microscopy.

Although not a normal part of quality control of concrete construction, it may become necessary to check that the concrete ingredients and their mix proportions are as specified. Such a requirement may arise if suspicions have been raised regarding an unauthorized change of materials, or from unsatisfactory concrete appearance or performance. Petrographic examination can be used to establish the type/source of the ingredients and the mix proportions can be checked by modal analysis.

Once fresh concrete is placed, it is typically compacted (by vibration) to eliminate entrapped air bubbles. This is important as there is an inverse relationship between the porosity and strength of concrete and the presence of entrapped voids will reduce strength. The characteristics of air voids and the assessment of compaction have been discussed on p. 92. Durability issues can arise from air voids that form beneath reinforcement bars or coarse aggregate particles (as entrapped air voids or water voids). Figure **188** shows a case of a concrete floor slab where air voids had formed below coarse aggregate

particles at the upper surface. Trafficking in service had caused the aggregate particles to collapse into the voids giving the appearance of pop-outs at the surface.

The concrete ingredients must be adequately mixed and unsegregated to ensure a uniform product. After placing there is a tendency for the concrete to segregate while in the plastic state, especially in poor-quality mixes lacking cohesion. Workmanship factors, such as allowing fresh concrete to flow along a form or excessive vibration, will exacerbate segregation. Segregation manifests itself in two main ways, which can both be identified and assessed by petrographic examination. The first consists of separation of coarse aggregate towards the bottom of the form, which can be detected in core samples (**189**). The second, which is characteristic of wet mixes, is the bleeding of mix water (and cement grout) to the surface. Bleeding can result in the formation of less durable, highly microporous zones and bleeding channels. Bleeding channels are readily observed in thin section using fluorescence microscopy, owing to their tortuous course (**190**) and associated enhanced microporosity (see **197**). An external manifestation of bleeding is the deposition of a cement-rich laitance layer on the concrete surface. As the laitance typically has a very high W/C, it is porous, soft, and weak. When excessive laitance is present on a concrete floor slab or pavement surface, it may consequently be prone to dusting.

Adequate curing and finishing are crucial to the production of durable concrete. The objectives of curing are to prevent the loss of moisture and to control the temperature of concrete for a period, to achieve the

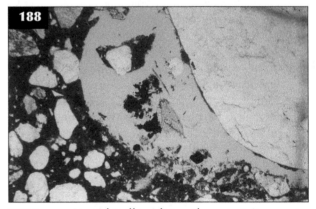

**188** Large air void (yellow) beneath a quartzite coarse aggregate particle (white) that was the cause of apparent 'pop-outs' on a concrete floor slab; PPT, ×35.

desired strength and permeability. Failure to control moisture loss at surfaces promotes bleeding, and along with exposure to low temperatures, these conditions can reduce concrete strength and durability. By contrast, excessively high curing temperatures can result in early thermal cracking and, in some cases, deleterious delayed ettringite formation (DEF).

One example of concrete finishing is that many concrete floors are constructed with dry shake surface finishes to improve abrasion resistance. Dry shake materials are factory prepared blends of cement, hard fine aggregate, and sometimes admixture and pigment, applied dry as a thin layer (2–3 mm) and trowelled into the fresh concrete base (191). Excessive trowelling when finishing concrete floors can mix dry shake toppings into the underlying base concrete, reducing their beneficial effect and giving a different appearance to that intended. Figure 192 shows an example of this from a warehouse ground floor slab where the variable appearance of the dry shake finish was unacceptable to the owner.

**189** Surface of a concrete core sample (150 × 100 mm area) exhibiting segregation of coarse aggregate (dark grey) from the cement matrix (light grey).

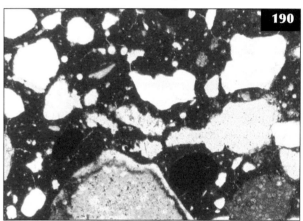

**190** Concrete exhibiting a bleeding channel (yellow) running from left to right; PPT, ×35.

**191** Finely ground slice (100 × 75 mm) through the top of a concrete floor slab, showing the 2–3 mm dry shake topping (dark grey, upper).

**192** Top surface of six 100 mm diameter core samples taken from a concrete floor slab with a variable dry shake surface finish.

# EXAMINING DETERIORATED AND DAMAGED CONCRETE

## OVERVIEW

Although most concrete is durable and will achieve good design life service, it can sometimes suffer from distress. Dense, good-quality concrete is highly resistant to weathering owing to its low permeability. However, most external concrete surfaces will eventually show evidence of deterioration caused by weathering mechanisms. The type of mechanisms operating will depend on the environment of use and may variously include driving and/or acid rain, cyclic wetting and drying, thermal cycling, leaching, freeze–thaw mechanisms, and salt crystallization. Examination of the microscopic appearance and textures of concrete gives a good indication of its condition and enables screening for evidence of deterioration. Concrete can sometimes suffer from deterioration caused by deleterious reactions or attack by aggressive agents. The petrographical study of crack and microcrack patterns, mineralogical changes, and secondary deposits provides crucial diagnostic information for determining causes of concrete distress. When investigating concrete deterioration it is important to establish whether the mechanisms of deterioration are intrinsic or extrinsic, as this will have important implications for repair and remediation of the affected concrete structure. Intrinsic mechanisms (e.g. AAR) are essentially internal reactions involving deleterious substances cast into the concrete at the time of construction and, consequently, can affect the whole section of the concrete member. In contrast, extrinsic mechanisms (e.g. the thaumasite form of sulfate attack [TSA]) involve agents from outside the concrete and produce a zone of deterioration confined to near the outer surface.

## CARBONATION AND REINFORCEMENT CORROSION

Carbonation of concrete is a reaction between carbon dioxide ($CO_2$) from the atmosphere and the various components of the hardened cement matrix to form carbonate minerals, as follows:

| Portland cement matrix component | Carbonation reaction products |
| --- | --- |
| Calcium hydroxide | Calcium carbonate and water |
| Calcium silicate hydrate | Calcium carbonate, silica gel, and water |
| Calcium aluminate hydrate | Calcium carbonate, alumina gel, and water |
| Hydrated ferrite phases | Calcium carbonate, ferric oxide, alumina gel, and water |
| Ettringite and calcium monosulfate | Gypsum, alumina gel, and water |

The most significant reaction is that involving the reaction of calcium hydroxide to calcium carbonate and therefore carbonation of concrete is generally summarized as:

$$Ca(OH)_2 \; + \; CO_2 \; \rightarrow \; CaCO_3 \; + \; H_2O$$
Calcium hydroxide · · · · · · Calcium carbonate

Carbonation usually advances inwards from the exposed concrete surfaces to form a carbonated layer (with a carbonation front). It may also advance along cracks or be associated with porous aggregate particles. The rate of carbonation depends on the environmental conditions and permeability of the concrete and can be predicted by modelling if these factors are known. The environmental conditions likely to promote high rates of carbonation are relative humidity in the 50–75% range, high carbon dioxide concentrations, and high temperature. Carbonation has important implications for the durability of concrete, notably reinforcement corrosion. Carbonation causes the alkaline protection of steel reinforcement bars to be lost by lowering the pH from 13–14 in uncarbonated concrete to 8.6 in fully carbonated concrete. In damp conditions this can cause deleterious corrosion of steel reinforcement.

Depth of carbonation is commonly assessed by spraying freshly broken concrete surfaces with phenolphthalein indicator solution (phenolphthalein in a mixture of ethanol and water), which stains concrete of >pH 9 purple. This indirect staining method has the advantage of being quick and inexpensive. However, it also has the disadvantage of detecting only fully carbonated cement paste, and as partially carbonated areas are missed, tends to underestimate the maximum depth of carbonation (St John et al., 1998). Direct optical microscopical examination (in thin section) is the definitive method for determination of carbonation depth as both fully and partially carbonated cement paste are readily observed. Depth of carbonation in relation to the outer surface and the position of reinforcement are routinely assessed during petrographic examination of hardened concrete.

In thin section, carbonated cement paste is readily detected by the presence of clumps of calcium carbonate crystals. Calcium carbonate ($CaCO_3$) has three polymorphic forms, these being calcite, aragonite, and vaterite. Calcite, the most stable, is the ultimate form found in carbonated concrete. In terms of optical properties, calcite is extremely birefringent, exhibiting pale high-order interference colours in cross-polarized light. Figure 193 shows the appearance of carbonated cement matrix in comparison with an uncarbonated area. For a particular part of the cement matrix the degree of carbonation can be described thus:

- Complete – no residual cement matrix other than occasional unreacted relics.
- Partial – evidence of mixed carbonate crystallites with isotropic matrix.
- Faint – occasional carbonate crystallites, <25% of the area.

In addition to the carbonation front advancing from the outer surface, carbonation can also spread along cracks (194). In terms of durability, this may move the effective depth of carbonation forward from the general carbonation front. Fine cracks of, for instance, 60 μm width or greater may allow aggressive agents ($CO_2$, moisture, chlorides) to reach the reinforcement long before the carbonation front has progressed through the cover concrete. Based on the fact that depth of carbonation increases with time, attempts have been made to use carbonation along cracks in concrete as an indicator of crack age. However, as numerous other factors affect the rate of carbonation such estimations of crack age are at best qualitative and should be regarded as unreliable (Jana & Erlin, 2007).

Steel reinforcement bars are commonly incorporated within concrete elements to increase the tensile strength.

However, this creates a durability issue, as steel will corrode in carbonated or chloride-exposed concrete. Reinforcement corrosion is now the greatest cause of deterioration of concrete structures, because designers and specifiers of works constructed in earlier decades were reliant on codes that did not fully take account of this phenomenon.

The corrosion of steel is an electrochemical process that is initiated by a difference in electrical potential along the steel in reinforced concrete. An electrochemical cell is set up with anodic and cathodic regions, connected by an electrolyte in the form of pore water in the hardened cement matrix. At the anode, positively charged ferrous ions pass into solution while negatively charged free electrons pass through the steel into the cathode and combine with water and oxygen to form hydroxyl ions. These travel through the electrolyte and combine with the ferrous ions to form hydrated iron oxides (ferrous hydroxide and ferric hydroxide), which are converted by further oxidation to rust. The reactions involved may be expressed as:

- Anodic reaction:
$$3Fe + 8OH^- \rightarrow Fe_3O_4 + 4H_2O + 8e^-$$
- Cathodic reaction:
$$8e^- + 4H_2O + 2O_2 \rightarrow 8OH^-$$

**193** Concrete with areas of both carbonated (brown) and uncarbonated (black) cement matrix. Flint coarse aggregate particles appear grey/speckled and quartz fine aggregate particles are shown grey/white; XPT, ×35.

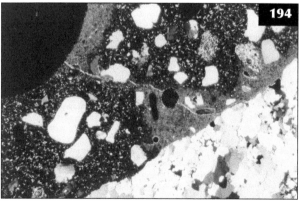

**194** Carbonation spreading along a fine crack (yellow) in the cement matrix. Carbonated cement matrix (mainly calcite) appears brown, while uncarbonated cement appears black and exhibits brightly coloured specks of portlandite. A small air void is seen upper left; XPT, ×35.

The resulting rust occupies a larger volume than the parent material and the expansion resulting from rusting of reinforcement causes the surrounding concrete to crack and spall, the bond between the concrete and reinforcement is lost and steel reinforcement is weakened through loss of section. Cracks caused by reinforcement corrosion may be seen over the bar position as linear cracks or surface spalling, or connect between bars to delaminate the cover concrete. The cracks may be filled by rust deposits (**195**).

Differences in electrical potential that may cause corrosion can arise from differences in the environment along a reinforced concrete element. These include variations of exposure to moisture, oxygen, and salts, differences in the depth of cover concrete, or where two dissimilar metals are connected. Carbonation is a common cause of reinforcement corrosion for the reasons discussed earlier in this section. Another common cause of corrosion is exposure to chloride ions and in concrete these can originate from various internal or external sources. Potential internal sources include marine aggregates and calcium chloride accelerator ($CaCl_2$). Use of calcium chloride as an accelerator ceased in the United Kingdom in 1997, following imposition of limits on total chloride content in reinforced and prestressed concrete. External sources of chlorides include sea water (and sea spray), airborne chlorides in coastal areas, and de-icing salts used on highways. Chloride ions cause corrosion cells to form by activating surface locations of the steel to form an anode, while the remaining passivated surface acts as a cathode.

To estimate the risk of reinforcement corrosion in a concrete element, one approach would be to determine chloride content (by chemical analysis of dust or lump samples) and the depth of carbonation (by indicator solution or petrography). Then if the exposure conditions are known (dry, >60% relative humidity, or damp, 70–80% relative humidity), the charts in BRE Digest 444 (Building Research Establishment, 2000) can be used to estimate the risk on a scale that runs from negligible to extremely high. For example, uncarbonated concrete in a dry environment that has a chloride content of <0.4% (by mass of cement) would be regarded in general terms as being at negligible risk of reinforcement corrosion.

## CRACKING

When reinforced concrete is designed it is assumed that the concrete will crack owing to thermal and humidity cycles; however, by careful design and detailing, cracks can be controlled and crack widths limited (Metha & Monteiro, 2006). Concrete is liable to crack for a variety of other reasons that may affect the durability, structural integrity, watertightness, sound transmission, and aesthetics of the structure. Unexplained cracking is a common reason for engineering investigations of concrete structures.

The main causes of concrete cracking are listed in *Table 22* and further details can be found in Concrete Society Technical Report No. 22 (1992). It is not easy to distinguish between different crack formations. Often, a number of laboratory tests and compilation of the complete history of the project, including concrete mixture design, placement condition, curing methods, formwork removal, and loading history, is required (Metha & Monteiro, 2006).

Petrographic examination of concrete slices and thin sections can aid the diagnosis of the cause of cracking and the determination of its severity. Examination (in ultraviolet light) of samples impregnated with fluorescent resin is used to highlight crack systems, which aids observation of crack morphology. Crack properties are usually described manually but may also be determined using image analysis techniques (Litorowicz, 2005). As a minimum the petrographer would normally describe such crack features as width (and changes in width), orientation, and distribution, whether the cracks occur in aggregate or cement paste, details of the adjacent cement paste, and the presence/identity of cracking infillings. Cracks types may be classified based on their nominal width (*Table 23*).

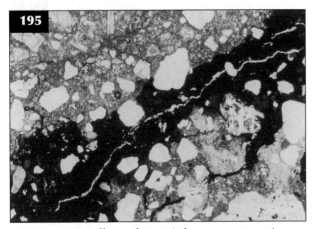

**195** Concrete suffering from reinforcement corrosion exhibiting a crack (yellow) lined with rust deposits (dark brown). Fine aggregate particles appear white and the cement matrix is shown light brown; PPT, ×35.

**Table 22** Causes of cracking in concrete (after Concrete Society, 1992)

| Time of crack | Crack type | Causes |
|---|---|---|
| After hardening | Physical | Shrinkable aggregates<br>Dying shrinkage<br>Crazing |
| | Chemical | Reinforcement corrosion<br>Alkali–aggregate reactions (AAR)<br>Delayed ettringite formation (DEF)<br>Cement carbonation |
| | Thermal | Freeze–thaw cycles<br>External seasonal temperature variations<br>Early thermal contraction |
| | Structural | Accidental overload<br>Creep<br>Design loads |
| Before hardening | | Early frost damage |
| | Plastic | Plastic shrinkage<br>Plastic settlement |
| | Constructional movement | Formwork movement<br>Subgrade movement |

**Table 23** Classification of cracks

| Crack type | Nominal width | Typical length |
|---|---|---|
| Fine microcracks | <1 µm | A few millimetres |
| Microcracks | 1–10 µm | 1–30 mm |
| Fine cracks | 10–100 µm | Up to 300 mm |
| Cracks | 100 µm–1 mm | Up to several metres |
| Large cracks | >1 mm wide | Up to several metres |

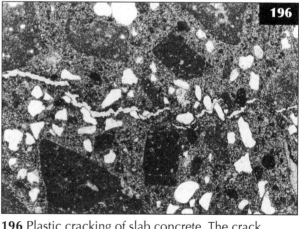

**196** Plastic cracking of slab concrete. The crack appears as a series of tension gashes (yellow); PPT, ×35.

Cracks that occur in fresh concrete include plastic cracks. Plastic shrinkage cracks are caused by excess bleeding and plastic settlement cracks are usually attributed to rapid early drying. In thin section, plastic cracks may appear as a linear series of tension gashes that run through the cement matrix and around aggregate particles (**196**). Another type of early cracking is known as 'crazing', which can occur on slab and wall surfaces as an irregular network of fine cracks (often with close spacing).

In hardened concrete, cracks running perpendicular to the surface (slabs and walls) may be caused by early age thermal movement or drying shrinkage. A certain amount of microcracking of the cement matrix is present in all concrete and is considered to be a normal concrete feature. The amount of microcracking caused by drying shrinkage is related to the W/C of the concrete, with more being present at higher W/C. Figure **197** shows a shrinkage microcrack within a concrete bleeding channel with localized high W/C. Microcracks generated by shrinkage tend to meet at triple junctions in the paste, to radiate from aggregate surfaces, run along parts of aggregate surfaces, and initiate on voids (French, 1991a). In reinforced concrete, expansion due to ASR or DEF causes patterns of fine cracks on concrete surfaces, along with cracking and microcracking at depth. The various cracks may be filled with secondary deposits of alkali–silica gel or sulfate minerals that may also exude on to concrete surfaces. Reinforcement corrosion causes cracking, spalling, and delamination of concrete often with associated brown rust deposits. Crack patterns associated with deleterious reactions are discussed later in their respective sections. Cracking caused by structural movement often runs through the aggregate particle as well as the cement matrix. Figure **198** shows concrete from a concrete cladding panel that cracked due to flexural loading of the panel in service.

## WEATHERING BY LEACHING, FROST ATTACK, AND SALT CRYSTALLIZATION

Dense, good-quality concrete is resistant to weathering owing to its low permeability. However, most external surfaces will eventually show evidence of deterioration caused by natural weathering processes such as cyclic wetting and drying, leaching, freeze–thaw mechanisms, and salt crystallization.

Leaching of the cement matrix occurs when moisture is able to pass through concrete. Calcium hydroxide (portlandite) is leached from cement hydrates and may be deposited within air voids or cracks (**199**). Consequently, the leached cement matrix may appear depleted of portlandite crystallites. Ettringite may also be leaching from the cement matrix to line voids and cracks (**200**). Abundant secondary deposits of ettringite associated with cracks may be symptomatic of sulfate attack of the aluminate phases, rather than just leaching of existing ettringite. The importance of moisture penetrating through concrete is frequently underestimated as a cause of damage to concrete (French, 1991a). Major leaching will cause considerable weakening of the cement matrix. Moisture is also required for natural weathering processes such as frost attack and salt crystallization and promotes deleterious reactions such as sulfate attack and AAR.

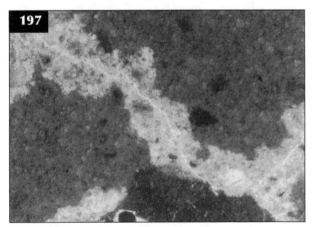

**197** Close view of a microcrack (yellow) in the cement matrix of structural concrete. The microcrack has formed in a zone of high microporosity associated with a bleeding channel (light green); UV, ×150.

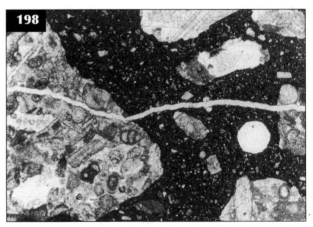

**198** Structural cracking of a concrete cladding panel. A crack (yellow) runs both through the cement matrix (dark brown) and limestone coarse aggregate particles (light brown); PPT, ×35.

Where leaching solutions run over concrete surfaces, deposits of white 'lime' can occur. The lime is built up in layers, which may become carbonated soon after deposition to form calcium carbonate (calcite). When in contact with the ground, layered calcite deposits may be stained brown by iron compounds (201). In certain circumstances a considerable thickness of calcite can be deposited and features such as stalactites, stalagmites, and calcareous tufa can be formed. Secondary deposits of lime have been known to block drains after being leached from cementitious materials such as bedding screeds for paving (202).

Popcorn calcite deposition (PCD), also known as cornflake calcite deposition or bicarbonation, is a form of calcite deposition found within deteriorated cement pastes of cementitious materials. PCD has been found associated with a wide range of different deterioration mechanisms and deleterious reactions, all of which involve considerable leaching. In thin section, the texture of PCD comprises tight clusters of calcite crystals (rosettes), which develop relatively evenly within an otherwise decalcified cement matrix that consists largely of silica gel (Sibbick & Crammond, 2003).

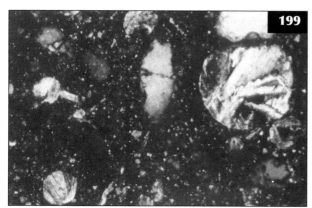

**199** Concrete subjected to leaching with secondary deposits of portlandite (brightly coloured) filling four small air voids. The cement matrix appears portlandite depleted (black); XPT, ×150.

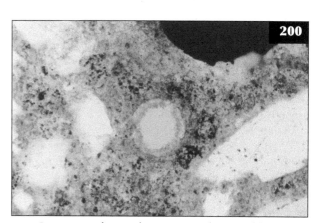

**200** Concrete subjected to a minor degree of leaching with acicular secondary deposits of ettringite lining a small air void (centre); PPT, ×150.

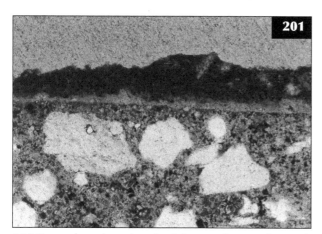

**201** Secondary deposits of calcium carbonate (brown) on the outer surface of leached concrete. In this example the deposits have been stained brown by iron compounds; PPT, ×150.

**202** Extensive secondary deposits of calcareous tufa found blocking the drains of a shopping centre, consisting of porous layers of calcium carbonate (pale pink); XPT, ×75.

Wet concrete can be damaged by freeze–thaw cycles in temperate and cold climates. Frost attack causes irregular cracks to form subparallel to the exposed concrete surfaces or corners. The cracks normally run through the cement matrix and around coarse aggregate particles. Petrographically, cracking is the main feature observed but the concrete may also exhibit some degree of leaching. There is another type of cracking induced by freeze–thaw cycles, termed 'D-cracking' that is linked to the coarse aggregate. Coarse aggregates from certain sedimentary rock sources (both natural gravel and crushed rock) are susceptible owing to the characteristics of their pore structure. D-cracking manifests itself after 10–15 years of exposure, as fine cracks near the edges (and joints) of slabs, which run through both the coarse aggregate and the cement matrix, parallel to the edges/joints.

Frost has undesirable effects on fresh concrete. If concreting has been undertaken in cold weather, slower than normal hydration may lead to the formation of a relatively porous cement matrix, with a lighter than normal colour for the given W/C. Placing concreting in freezing conditions can cause the concrete not to set properly, resulting in a weak and friable product.

Concrete that is exposed to salts is prone to salt weathering. Salts may originate naturally from sea water, sea spray, coastal air, ground/groundwater or be applied by man as de-icer salts. Salt crystallization causes flaking and scaling of concrete surfaces up to several millimetres at a time, often associated with white salt deposits. During sample collection and preparation care must be taken not to lose soluble salts from contact with water.

## SULFATE ATTACK FROM GROUNDWATER

Sulfate attack is a term used to describe a series of deleterious chemical reactions between sulfate ions and the components of hardened concrete, principally the cement matrix, caused by exposure of concrete to sulfates and moisture (Skalny et al., 2002). The sulfates of greatest concern for the durability of concrete are salts found in natural soils and groundwaters such as sulfates of sodium, potassium, magnesium, and calcium. In addition, groundwater that has been contaminated with fertilizer or industrial effluent may also contain ammonium sulfate. The mechanisms of sulfate attack are complicated, involving a number of overlapping chemical reactions, which are not yet fully understood. However, it is known that the extent of attack depends on the amount of sulfate in solution and that the aggressiveness of the sulfate salts is broadly related to their solubility. The solubility of sulfate salts running from most to least is: ammonium, magnesium, sodium, potassium, and calcium. Calcium sulfate (found in gypsum-bearing soils) attacks only the

calcium aluminate phase of the cement matrix to form ettringite (calcium sulfoaluminate) as follows.

Calcium aluminate + $CaSO_4.2H_2O \rightarrow 3CaO.Al_2O_3.3CaSO_4.32H_2O$ hydrates

The other sulfate salts (ammonium, magnesium, sodium, potassium) react with calcium hydroxide (portlandite) in the concrete to form calcium sulfate (gypsum) as shown below. This calcium sulfate may then interact with the calcium aluminate phase to form ettringite, as in the reaction detailed above (St John et al., 1998).

$$(NH_4)_2SO_4 + Ca(OH)_2 + 2H_2O \rightarrow CaSO_4.2H_2O + 2NH_3$$
$$MgSO_4 + Ca(OH)_2 + 2H_2O \rightarrow CaSO_4.2H_2O + Mg(OH)_2$$
$$Na_2SO_4 + Ca(OH)_2 + 2H_2O \rightarrow CaSO_4.2H_2O + 2NaOH$$
$$K_2SO_4 + Ca(OH)_2 + 2H_2O \rightarrow CaSO_4.2H_2O + 2KOH$$

Magnesium sulfate also attacks the calcium silicate phases of the cement matrix to form gypsum, brucite (magnesium hydroxide), and hydrated silica. Below a pH of 10.6, it may also attack ettringite to form more gypsum, brucite, and hydrated alumina (St John et al., 1998).

Sulfate attack from an external source of sulfates (such as groundwater) will exhibit a zone of deterioration that works inwards from the surfaces exposed to the sulfates. Deterioration manifests itself, firstly, by causing cracking associated with the expansive formation of ettringite and/or gypsum and, secondly, by the dissolution and weakening of the cement matrix. Petrographically, the cracks and microcracks will be observed in thin section along with secondary deposits of sulfate minerals (usually ettringite or gypsum) filling cracks and air voids. Figure 203 shows concrete from a pile cap of a flyover that exhibits evidence of sulfate attack consisting of a network of microcracks and air voids filled with ettringite. Ettringite is identified from its small, needle-like crystals, first-order grey interference colours, and its colourless appearance in plane-polarized light. Fluorescent microscopy can be used to highlight the cracks produced by deterioration and Figure 204 shows this for an example of a pile from a building suffering from external sulfate attack.

Guidance regarding the design of buried concrete elements that may be exposed to sulfates of other aggressive ground conditions is given in BRE Special Digest 1 (Building Research Establishment, 2005).

## THE THAUMASITE FORM OF SULFATE ATTACK (TSA)

TSA is a special form of external sulfate attack that can lead to particularly severe deterioration in buried concrete. In conventional sulfate attack, incoming sulfate ions react with the calcium aluminate phases and calcium hydroxide in the cement matrix to form either ettringite or gypsum. TSA differs from conventional sulfate attack in that the C-S-H of the cement matrix are attacked by external sulfates to form thaumasite ($CaSiO_3.CaCO_3.CaSO_4.15H_2O$). As the main C-S-H binding phase of the cement is attacked, TSA can be more severe than conventional sulfate attack and, importantly, SRPC offers no protection. In addition, limestone aggregate particles are also consumed by TSA.

Thaumasite is believed to form primarily at temperatures <5–10°C, although evidence is mounting that it may form even at elevated temperatures (Skalny *et al.*, 2002). The replacement of C-S-H by thaumasite results in softening of the cement matrix into a white, mushy incoherent mass. In buried concrete elements, TSA has been found penetrating the outer 50 mm and exhibiting four distinct zones of gradually decreasing degradation inwards (Sibbick & Crammond, 1999). The following zones can be recognized for TSA in hand specimen and thin section:

- *Zone 1* (outer) – cement paste almost completely replaced by thaumasite leaving aggregate particles embedded in soft white mush.
- *Zone 2* (outer middle) – abundant cracks and large cracks infilled with thaumasite running subparallel to the outer surface. Haloes of thaumasite around coarse and fine aggregate particles.
- *Zone 3* (inner middle) – fine thaumasite-filled cracks running subparallel to the outer surface.
- *Zone 4* (inner) – no evidence of attack in hand specimen. Needles of thaumasite and sometimes ettringite filling voids, seen in thin section.

Figure 205 illustrates the typical appearance of TSA zone 3 in thin section. Ettringite and thaumasite cannot always be distinguished from examination of thin sections alone, and if TSA is suspected, optical microscopy should always

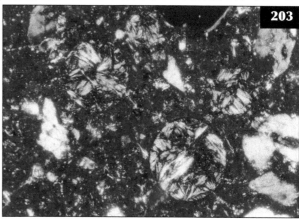

**203** Close view of concrete suffering from external sulfate attack, showing small air voids and microcracks packed with acicular ettringite crystals (grey). The uncarbonated cement matrix appears portlandite depleted (black); XPT, ×150.

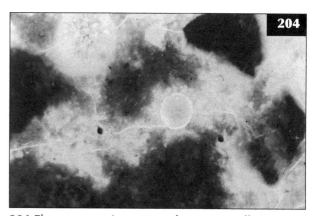

**204** Fluorescent microscopy of concrete suffering from external sulfate attack, showing microcracks (yellow) running through the cement matrix and connecting with ettringite-filled air voids; UV, ×150.

**205** Zone 3 TSA showing fine cracks running subparallel to the outer surface filled with thaumasite (orange); XPT, ×35.

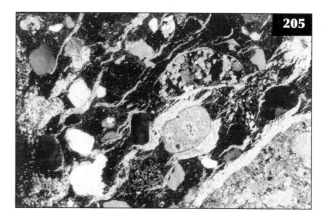

be supplemented by EPM and/or XRD (Eden, 2003). A comparison of the chemical compositions of ettringite and thaumasite follows (normalized to 100% and excluding $CO_2$ and $H_2O$) (*Table 24*).

In terms of optical properties, ettringite and thaumasite are very much alike, both forming small, needle-like crystals. The most distinguishing features of thaumasite are that it often (but not always) has a yellow colour (light straw yellow to bright yellow) in plane-polarized light (**206**), rather than being colourless like ettringite. For this reason and to avoid confusion, it is best not to impregnate samples for TSA investigation with fluorescent yellow-dyed resin (colourless or blue-dyed resins are suitable alternatives). The other distinguishing optical characteristic is that thaumasite

usually has a significantly higher birefringence than ettringite. The interference colours of ettringite are rarely above first-order grey, while thaumasite is typically first-order yellow to second-order yellow (**207**).

Guidance regarding the investigation and prevention of TSA is given in: *The Thaumasite Form of Sulfate Attack: Risks, Diagnosis, Remedial Works and Guidance on New Construction* (DETR, 1999).

## INTERNAL SULFATE ATTACK (INCLUDING DELAYED ETTRINGITE FORMATION)

Internal sulfate attack on concrete is due either to the inadvertent inclusion of materials containing sulfate, or aggregates that can oxidize to produce sulfate or the delayed formation of ettringite (St John *et al.*, 1998). The

**Table 24** Comparison of the chemical compositions of ettringite and thaumasite

|  | Ettringite | Thaumasite |
|---|---|---|
|  | $3CaO.Al_2O_3.3CaSO_4.31H_2O$ | $CaSiO_3.CaCO_3.CaSO_4.15H_2O$ |
| $SiO_2$ | 0 | 19% |
| $Al_2O_3$ | 15% | 0 |
| CaO | 50% | 55% |
| $SO_3$ | 36% | 26% |

**206** Close view of thaumasite in plane-polarized light showing a mass of needle-like crystals with straw yellow colour; ×150.

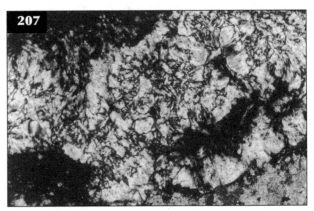

**207** Close view of thaumasite in cross-polarized light showing a mass of needle-like crystals with first- and second-order interference colours (grey/yellow); ×150.

most likely cause of internal sulfate attack is the use of sulfate- or sulfide-contaminated aggregates. These may release sulfate ions within the concrete that may react with the cement matrix to form ettringite and/or gypsum as described on p. 102. Although less common, internal sulfate attack can result from cement or mixing water with high sulfate contents. In thin section, the appearance of concrete suffering from internal sulfate attack includes the occurence of a network of cement matrix cracks, that along with air voids, are filled with secondary deposits of ettringite or gypsum. Unlike external sulfate attack, evidence of deterioration will not be confined to the surfaces.

Internal sulfate attack is responsible for 'the mundic problem' in southwest England. Here, the expansive cracking of concrete blocks and mass concrete has been caused by the use of mining waste aggregate containing pyrite ($FeS_2$). The mechanisms of deterioration involve oxidation of the pyrite (Bromley & Sibbick, 1999). Consequently the equity value and resale potential of many, mainly pre-1950s domestic properties have been adversely affected by uncertainties about the nature of the concrete used in their original construction. A petrography-based sampling and testing procedure (described in Stimson, 1997) was therefore adopted to determine if buildings contained concrete with potentially expansive aggregate (and the likely risk of future deterioration). The procedure involves taking representative core samples (50 mm diameter) from the concrete elements of the building and examining them in hand specimen using a low-power microscope (and in thin section if uncertainties arise). This screens for evidence of deterioration and is used to classify the aggregate lithologies present into either Group 1 (inert) or Group 2 (potentially deleterious). A detailed atlas of concrete aggregates used in southwest England has been collated by Bromley (2002). Figure 208 shows concrete blocks made with china clay waste, which is a commonly seen Group 1 aggregate. Figure 209 shows concrete made with potentially deleterious crushed metasedimentary mining waste aggregate, which is classified as Group 2. Providing that the concrete appears sound and contains less than 30% of Group 2 aggregate then the property is likely to be considered safe for mortgage purposes by lenders. Since its introduction, the petrographic assessment procedure has enabled approximately 80% of previously blighted houses in southwest England to be saleable again.

Delayed ettringite formation (DEF) is a rare but potentially severe form of internal sulfate attack of concrete (Quillin, 2001). A certain amount of ettringite (hydrous calcium sulfoaluminate) forms as part of the normal cement hydration products. However, if the concrete is subjected to elevated temperatures during curing (>65°C) this ettringite formation may be 'delayed' and if the hardened concrete is subsequently exposed to moisture, the ettringite may re-form causing expansion and cracking, months or years after construction. Concrete may be exposed to high temperatures during curing as a result of

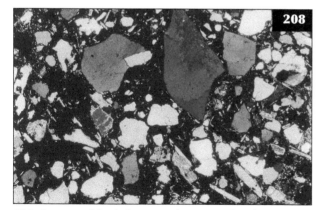

**208** Concrete block from southwest England containing Group 1/1 china clay waste aggregate, consisting chiefly of quartz (grey), with minor feldspar (grey), muscovite mica (brightly coloured), and tourmaline (orange). The uncarbonated cement matrix appears black; XPT, ×35.

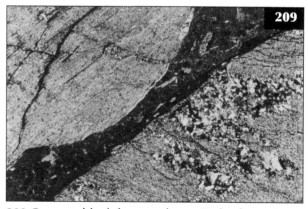

**209** Concrete block from southwest England containing Group 2/1 crushed phyllite ('killas') aggregate. The aggregate particles are coated by iron oxide (dark red) and the cement matrix is cracked (yellow) suggesting deterioration due to pyrite oxidation. The carbonated cement matrix appears brown; XPT, ×35.

steam curing (precast units), large pours (mass concrete or sections >0.5 m in size) or high ambient temperatures (in tropical climates). The use of certain mineral additions such as GGBS or PFA helps to reduce the peak curing temperature and prevent DEF. In reinforced concrete suffering from DEF, the outer surface will often exhibit map cracking while subparallel cracks form at depth. In unreinforced concrete, large cracks (up to 20 mm in width) may form both subperpendicular and subparallel to the outer surfaces. In thin section, concrete suffering from DEF exhibits distinctive partings around aggregate particles and microcracks in the cement paste that may both be filled with ettringite (Eden *et al.*, 2007). Figures **210** and **211** show the appearance of concrete from a large *in situ* concrete pour suffering from DEF. DEF is diagnosed by the presence of partings around aggregate particles (that usually become wider as particle size increases), the absence of an external source of sulfate, and a high temperature curing history. DEF may occur in combination with other deleterious reactions such as ASR, in which case it may be difficult to determine what the dominant mechanism of deterioration is (Thomas *et al.*, 2007).

### SEA WATER ATTACK
Concrete in marine environments is subject to deterioration from chemical attack, abrasion, freeze–thaw mechanisms, salt crystallization/scaling, and, if the concrete is reinforced, chloride-induced corrosion of steel reinforcement. The relative significance of each form of attack will depend on the exposure conditions of the structure, with the intertidal and splash zones usually being most at risk. Only chemical attack by sea water will be considered here as the other mechanisms are discussed in other sections.

Open sea water has a pH of 7.5–8.4 and contains 3.1–3.8% soluble salts (by weight) consisting (approximately) of 2% chloride, 1.1% sodium, 0.28% sulfate, 0.14% magnesium, 0.05% calcium, and 0.04% potassium. In seas that are partially enclosed, such as the Arabian Gulf, salt concentrations can rise considerably (5–10%). Chemical attack of concrete is mainly due to the presence in sea water of magnesium sulfate (Lea, 1970), which acts on the cement matrix. Magnesium sulfate reacts with the free calcium hydroxide (portlandite) to form calcium sulfate (gypsum) and precipitating magnesium hydroxide (brucite) as follows:

$$Ca(OH)_2 + MgSO_4.7H_2O \rightarrow CaSO_4.2H_2O + Mg(OH)_2 + 5H_2O$$

The magnesium sulfate also reacts with the calcium silicate hydrates of the cement matrix to form more gypsum, brucite, and hydrated silica as follows:

$$3CaO.2SiO_2.nH_2O + 3MgSO_4.7H_2O \rightarrow CaSO_4.2H_2O + 3Mg(OH)_2 + 2SiO_2.nH_2O$$

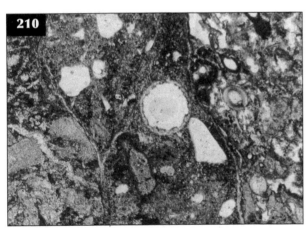

**210** Concrete suffering from DEF showing ettringite-filled partings around coarse aggregate particles and ettringite deposits lining a small air void; PPT, ×35.

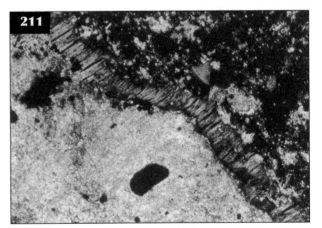

**211** Close view of concrete suffering from DEF showing an ettringite-filled (grey) parting around a limestone coarse aggregate particle (pale pink). The ettringite crystals are aligned perpendicular to the crack; XPT, ×300.

In good-quality concrete, the brucite precipitates in the pores at the surface forming a protective surface layer that impedes further reaction. Some precipitated calcium carbonate, $CaCO_3$ (as aragonite), arising from the reaction of calcium hydroxide and carbon dioxide ($CO_2$), may also be present blocking the surface pores (Neville, 1995). In addition, secondary deposits of ettringite (calcium sulfoaluminate) may be generated by sulfate attack of the C-S-H of the cement matrix (sulfate attack is discussed on p. 102).

Sea water causes the development of a zone of alteration adjacent to the outer surface, which will widen with time provided that the surface is not removed by abrasion. Petrographically, the zone of alteration and the occurrence of sea water attack can be detected by the presence of disruption and deposition of secondary minerals (brucite, aragonite, gypsum, ettringite) within the cement matrix and air voids. In thin section, brucite is difficult to differentiate from calcium carbonate minerals (calcite and aragonite) as it often occurs as small crystallites with similar refractive indices and birefringence (St John *et al.*, 1998). Consequently, the petrographer may only report the presence of carbonate minerals unless the presence of brucite has been confirmed by chemical analysis (XRD or SEM microanalaysis). Figure 212 shows concrete that has been exposed to sea water, having secondary deposits of carbonate minerals in air voids. Where present, gypsum is easily identified in thin section by its low relief and weak birefringence (up to first-order pale grey) and fibrous or lath-shaped crystals (213). It should be borne in mind that secondary minerals such as gypsum and ettringite are relatively soluble in sea water and are liable to be leached away, while brucite is almost insoluble and will remain *in situ*.

Although sea water contains alkali metals (sodium and potassium) they are thought unlikely to cause deleterious ASR as they are in the form of salts, rather than the alkali hydroxides known to be involved in ASR (ASR is discussed on p. 109).

## ATTACK BY ACIDS AND ALKALIS

Concrete may be subjected to chemical attack in industrial and other aggressive environments (Plum & Hammersley, 1984). Petrographic examination is able to determine the nature and depth of damage to the concrete and, in favourable circumstances, may provide an approximate estimation of the rate of attack.

Acidic solutions are among the most aggressive to Portland cement concrete and include both mineral acids, such as sulfuric, hydrochloric, hydrofluoric, nitric, phosphoric, and carbonic, and organic acids, such as lactic, acetic, citric, formic, humic, and tannic. All of the cement matrix hydrates are susceptible to acid attack but calcium

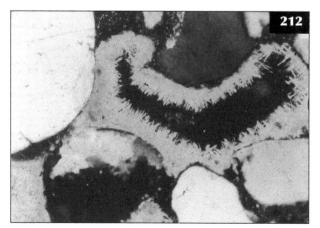

**212** Concrete exposed to sea water with secondary deposits of carbonate minerals, probably brucite and aragonite (light brown/yellow) in air voids. Quartz fine aggregate particles appear grey/white and the cement matrix is dark brown; XPT, ×150.

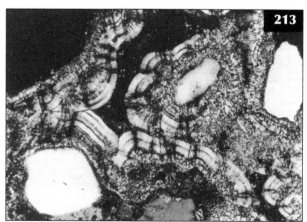

**213** Concrete suffering from sea water attack showing extensive replacement of the cement matrix by gypsum (grey). Quartz fine aggregate particles appear white/grey/black and air voids are shown dark green; XPT, ×150.

hydroxide (portlandite) is attacked most readily. Limestone aggregate will also be attacked by acid. The attacking acid can usually be identified by the presence of the salt of the acid deposited in the concrete. For example, sulfuric acid reacts with calcium hydroxide in the cement matrix to produce gypsum (calcium sulfate) as follows:

$$H_2SO_4 + Ca(OH)_2 \rightarrow CaSO_4.H_2O$$

As the attack proceeds all of the cement compounds are eventually broken down and leached away together with any calcareous aggregate. With sulfuric acid attack there is an additional issue with the gypsum formed by the initial reaction then reacting with the aluminate phases in the cement to form ettringite (calcium sulfoaluminate), which on crystallization can cause further expansive disruption of the concrete. Generally, for acid attack of concrete surfaces, the following characteristic zones can be recognized in thin section:

- *Zone 1* (outer) – exposed cement paste completely disintegrated together with any calcareous aggregate.
- *Zone 2* (middle) – portlandite depleted and acid salts deposited.
- *Zone 3* (inner) – unaffected.

Figure **214** shows an example of acetic acid attack of a suspended reinforced concrete floor slab at a factory involved in the production of cellulose acetate flake. The photograph shows the appearance of zone 2 following acid attack.

For sulfuric acid attack the following zones can be recognized in thin section:
- *Zone 1* (outer) – exposed cement paste completely disintegrated together with any calcareous aggregate.
- *Zone 2* (outer middle) – portlandite depleted and gypsum deposited.
- *Zone 3* (inner middle) – ettringite formed and sulfate attack.
- *Zone 4* (inner) – unaffected.

Another form of sulfuric acid attack is associated with the problem of hydrogen sulfide gas in sewers. Sewage effluent is normally alkaline and does not directly attack concrete sewer linings. However, sewage contains sulfur compounds that can be decomposed by anaerobic bacteria present in sewers to release hydrogen sulfide gas (Pomeroy, 1992). This gas is slightly heavier than air and is absorbed by the moisture coating the sewer walls which contain aerobic bacteria (Thiobacilli). The aerobic bacteria oxidize the hydrogen sulfide to sulfuric and sulfurous acids, which corrode vulnerable materials including concrete. The acid dissolves the cement matrix at concrete surfaces and reacts with it to produce expansive gypsum (calcium sulfate). In hand specimen, concrete surfaces exhibit exposed aggregate with the remaining outer 10–20 mm of cement paste being noticeably softened (like putty). Figure **215** shows the microscopical appearance of sewer concrete that has been subjected to this form of acid attack. The cement

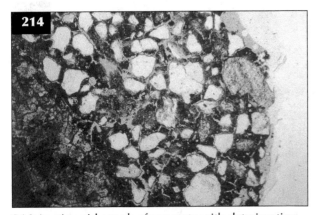

**214** Acetic acid attack of concrete with deterioration of the cement matrix (brown, with yellow fine cracks). Quartz fine aggregate particles are shown white and a dolomite coarse aggregate particle appears lower left; PPT, ×35.

**215** Sulfuric acid (from hydrogen sulfide gas) attack of precast concrete sewer lining, showing fine cracks filled with gypsum (white) and deteriorated cement matrix (light brown). Quartz fine aggregate particles appear white; PPT, ×35.

matrix appears more porous and fine cracks filled with gypsum run subparallel to the outer surface.

Uncarbonated Portland cement concrete is highly alkaline (pH >12.5) owing to the presence of calcium hydroxide ($Ca(OH)_2$) and alkali hydroxides (NaOH and KOH) in the hydrated cement. They are therefore relatively resistant to alkali attack. However, concentrated (>10%) alkaline solutions will attack the cement matrix of Portland cement concrete, with dissolution of the cement matrix typically occurring at a much slower rate than with acid attack. High-alumina cement (HAC) concrete is more susceptible to alkali attack than Portland cement concrete. Alkaline hydrolysis of HAC is discussed on p. 119. In addition to alkali attack, alkali-bearing solutions can cause AAR in concrete, which is discussed on p. 110.

## POP-OUTS

Pop-outs are conical cavities on concrete surfaces that range from a few millimetres to as much 100 mm in diameter and up to 50 mm in depth. They are formed by volume increase of aggregate particles or concrete contaminants immediately beneath the concrete surface. Pop-outs typically start as a semicircular cracks that, as expansion increases, become roughly circular with a cone of protruding concrete, which may be removed to form the cavity.

Expansive materials found to cause pop-outs include naturally occurring aggregate particles such as alkali-reactive particles, particles containing swelling clays, pyrite particles, organic matter, and particles susceptible to frost attack. Pop-outs have also been caused by artificial materials that have contaminated the concrete during production or transport, such as periclase, lime, or anhydrite (Katayama & Futagawa, 1997).

Pop-outs on concrete surfaces are not usually of structural significance but they do have implications for durability of reinforced concrete, as they locally reduce the depth of cover to the reinforcement. Most complaints about pop-outs are for aesthetic reasons as they are unsightly and certain types, such as pyrite (**216**) and organic matter (**217**), may produce a brown stain on the concrete surface.

## ALKALI–AGGREGATE REACTION (AAR)

AAR has been recognized as a rare, but potentially damaging form of concrete deterioration since 1940 (Stanton, 1940). There are two main types of deleterious AAR, the most common being alkali–silica reaction (ASR) and the second being the rather rare alkali–carbonate reaction (ACR). Both these types involve the interaction of reactive aggregate constituents with alkali (sodium and potassium) hydroxyl ions (from the cement or an external source) in the presence of moisture, to form deleterious reaction products (notably alkali–silica gel). The reactions often cause expansion and cracking that, while not directly causing structural failure, will reduce the service life of the concrete by allowing penetration by aggressive agents.

Successful diagnosis of AAR relies on the integrated assessment of all the evidence gleaned from documentary, site, and laboratory investigations.

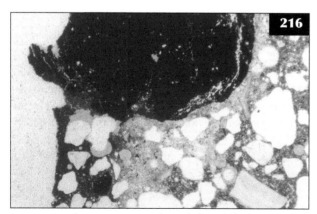

**216** View of a concrete surface (left) exhibiting a pop-out caused by a pyritic aggregate particle (black). The oxidized pyrite has stained the concrete surface brown (iron oxide); PPT, ×35.

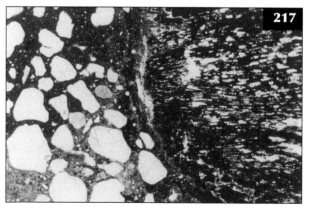

**217** Concrete contaminated with particles of lignite (black, right) that caused pop-outs on exposed concrete surfaces; PPT, ×35.

Almost all AAR investigations involve microscopical examination, which is the definitive method for identifying deleterious reactions. Guidance for investigating structures suspected of suffering from AAR can be found in Palmer (1992) and the structural effects are discussed in an Institution of Structural Engineers publication (1992). Guidance for avoiding AAR in new concrete construction can be found in BRE Digest 330 (Building Research Establishment, 2004) and Concrete Society Technical Report No.30 (1999).

ASR, the most common form of AAR, is a two-stage process. Initially alkali hydroxides react with susceptible siliceous aggregate constituents in the presence of moisture to form alkali–silica gel as follows (the chemical equation can also be written with potassium in place of sodium):

$$\underset{\text{Silica}}{SiO_2} + \underset{\text{Alkali}}{2NaOH} + \underset{\text{Water}}{H_2O} \rightarrow \underset{\text{Alkali–silica gel}}{Na_2SiO_3.2H_2O}$$

The gel can later absorb further moisture, swelling in the process and imposing expansive forces on the surrounding concrete matrix. These expanding reaction sites generate radiating microcracks and the gel migrates into the cracks. As ASR progresses, the reaction sites become linked by a network of widening cracks that may appear on the outer concrete surface as map cracking (218). Crack patterns and reaction sites can be successsfully investigated using a combination of polished slice and thin section examinations. Several schemes for quantifying petrographic AAR evidence have been proposed, including one by Sims *et al.* (1992). A classification for cracking associated with AAR as observed in thin section is provided in *Table 25*.

The relative ASR reactivity for different types of aggregate constituent is shown in *Table 15* and procedures for determining the potential alkali–silica reactivity of concrete aggregate combinations is discussed on pp. 69–73. Figure 219 shows the appearance in thin section of ASR in concrete made with crushed greywacke coarse aggregate, which is classified as being of 'high' reactivity in accordance with BRE Digest 330 (Building Research Establishment, 2004). Figure 220 shows ASR in concrete with flint (chert) coarse aggregate that is classified as being of 'normal' reactivity. In both examples alkali–silica gel is present, being characteristically amorphous and colourless in plane-polarized light (but may sometimes be yellow or brown). In cross-polarized light typical amorphous gel is isotropic. Some areas of gel may convert to a crystalline form that is birefringent with first-order white interference colours. Gel near surfaces may carbonate to give higher-order interference colours more typical of carbonate minerals. Several staining methods to aid the identification of alkali–silica gel are described in Appendix B. These are most useful for highlighting deposits of gel in field situations or on hand specimens as gel is usually readily identifiable in thin section.

Cement is the principal source of alkalis that may become involved in ASR. The alkali content of cement or concrete is determined by chemical analysis and is expressed as the $Na_2O$ equivalent ($Na_2O + 0.658K_2O$). Generally ASR does not occur unless the alkali content (as $Na_2Oeq$) of the concrete is >3.0 $kg/m^3$. Another possible internal source of alkalis is salt-contaminated aggregates (e.g. marine dredged). Also, in rare instances, aggregates that include certain alkali-containing

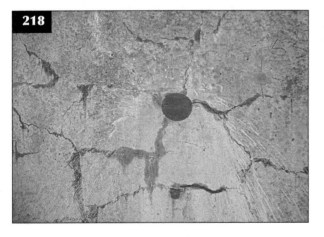

**218** View of a reinforced concrete wall exhibiting map cracking as a result of ASR. The scale is illustrated by an infilled core sample hole that is 100 mm in diameter.

**Table 25** Classification of crack severity associated with AAR

| Grade | Cracking classification | Microscopical observations |
|---|---|---|
| 0 | No cracking | No cracking observed |
| 1 | Very slight | Microcracks (1–10 μm wide), confined mainly to cement matrix and sporadically at cement/aggregate interfaces. No significant internal aggregate cracking |
| 2 | Slight | Microcracks and fine cracks up to 40 μm wide, usually isolated, not forming a network. May originate from aggregates but do not propagate far into surrounding cement matrix. Sporadic internal and peripheral cracking of aggregates may be present |
| 3 | Moderate | Fine cracks (10–100 μm wide) that begin to form an interconnecting network linking sites of expansive reaction. Internal and peripheral cracking of aggregates may be common. Cracks may contain secondary deposits (ASR gel, ettringite, and so on) |
| 4 | Severe | Cracks (100 μm–1 mm wide) form an interconnecting system of ASR gel-filled cracks that link reaction sites, throughout the concrete. Much internal aggregate cracking may be evident. May exhibit evidence of severe leaching with secondary deposits (ettringite, portlandite, calcite) in cracks |
| 5 | Very severe | Large cracks (>1 mm wide) abundant throughout. Much internal aggregate cracking may be evident. The effects of leaching may be very advanced |
| 6 | Concrete disintegrated | |

**219** ASR associated with crushed greywacke aggregate. A greywacke coarse aggregate particle (lower left) exhibits internal cracking (yellow) with cracks running into the cement matrix (brown) filled with alkali–silica gel (light grey); PPT, ×150.

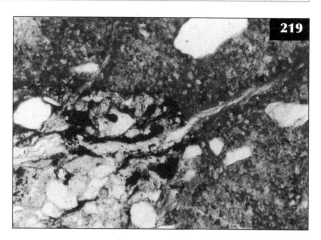

**220** ASR associated with flint (chert) coarse aggregate. A flint particle (dark brown) is internally cracked (yellow) and the adjacent cement matrix exhibits alkali–silica gel-filled cracks (white). The cracks interconnect with a small air void that is competely filled with gel (left); PPT, ×150.

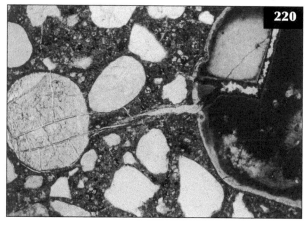

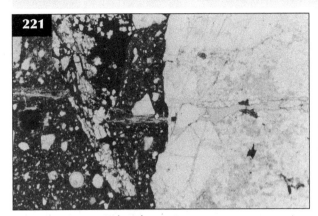

**221** ASR associated with granite coarse aggregate. A granite particle (right) exhibits internal cracking (yellow) and the crack continues into the surrounding concrete where it is seen to be filled with alkali–silica gel (light grey); PPT, ×35.

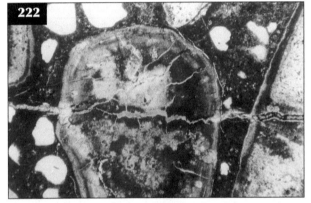

**222** ASR in concrete contributed to by de-icing salts. Flint (mottled brown/grey) aggregate particles exhibit internal cracking (blue) and cracks in the cement matrix are filled with alkali–silica gel (white); PPT, ×35.

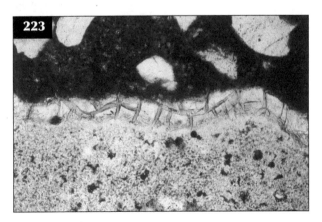

**223** Close view of alkali–silica gel (white) formed at the edge of a flint aggregate particle (light brown) in concrete suffering from ASR (related to de-icing salt exposure). The gel exhibits desiccation cracks (blue); PPT, ×150.

minerals (e.g. feldspars, phyllosilicates) have been found to release alkalis into concrete (Goguel & Milestone, 2007). Figure **221** shows ASR in concrete with granite coarse aggregate. During ASR, hydroxide ions may attack feldspars and micas in granite to release silica, alumina, and alkalis that contribute to continued ASR in the concrete (Yan *et al.*, 2004). An external source of alkali found in highway structures is de-icing salt (Katayama *et al.*, 2003). De-icing salt is halite (sodium chloride) and in solution this may react with portlandite in the cement paste to release alkali as follows:

$$2NaCl + Ca(OH)_2 \rightarrow 2NaOH + CaCl_2$$

Sodium chloride     Portlandite     Alkali

It is thought that rather than causing ASR, alkali from de-icing salts accelerates the reaction once it has already been initiated by alkali from inside the concrete (West, 1996). Figures **222** and **223** show concrete suffering from ASR that has been contributed to by de-icing salts. This particular example is described in a publication by Dunster (2002a, Case Study D) and involves crosshead beams of an elevated link road, which exhibited ASR only on outer surfaces and in drains that had been heavily contaminated with de-icing salt. The reactive aggregate constituent was flint (chert), of 'normal' reactivity. A combination of petrography and chemical analysis was used to determine that ASR had only occurred in areas of the concrete where the alkali content was >5 kg/m$^3$ Na$_2$Oeq.

Limestone and other carbonate rocks are commonly used as concrete aggregates. Three different AARs have been observed in carbonate rock aggregates:

1. ACR of dolomitic limestones that results in dedolomitization.
2. ACR of nondolomitic limestones that produces reaction rims.
3. ASR of various carbonate rocks containing reactive silica.

ACR is a rare and poorly understood form of AAR. The mechanism is thought to be different from ASR, though the associated concrete deterioration is often similar. Researchers have struggled to explain the mechanism of ACR expansion but most hypotheses centre on the possible role of dedolomitization. It is thought that alkali solutions react with the mineral dolomite (in the aggregate) to form brucite and calcite as follows:

$$CaMg(CO_3)_2 + 2NaOH \rightarrow Mg(OH)_2 + CaCO_3 + Na_2CO_3$$
Dolomite      Alkali      Brucite      Calcite

While there is no doubt that the dedolomitization reaction does occur in concrete with carbonate aggregate, whether it is responsible for deleterious expansion has not been established. In many cases of apparent ACR, it has been shown that expansion has in fact been caused by ASR involving cryptocrystalline silica within carbonate aggregates. The microscopical textures associated with ACR and dedolomitization are illustrated in Katayama (2003). Figures **224** and **225** show an example of ASR within concrete with limestone coarse aggregate, where the silica for the reaction was provided by siliceous bands (chert) within the limestone.

Diagnosing the cause(s) of deterioration in concrete is complicated by the common coexistence of evidence for several different mechanisms. For example, coexistence of the following combinations of concrete deterioration mechanisms has been reported in case studies by Sims *et al.* (2004):

- ASR and leaching.
- ASR and frost attack.
- ASR (dominant) and DEF.
- DEF (dominant) and ASR.
- TSA (dominant) and ASR.
- Acid attack, reinforcement corrosion, and ASR.

Figure **226** shows pile cap concrete from a bridge pier suffering from map cracking. In thin section, evidence for both DEF and ASR was apparent. In this case, DEF appeared to be dominant as a high abundance of ettringite-filled partings and cracks suggested that much more expansion was associated with DEF than ASR.

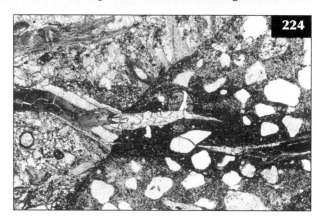

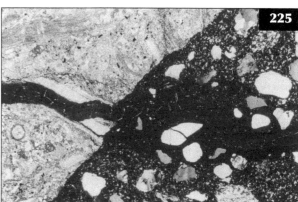

**224, 225** Concrete with ASR from siliceous limestone coarse aggregate (upper left) seen in PPT (**224**) and XPT (**225**). A fine crack originates in the coarse aggregate particle and runs through the surrounding cement matrix. The crack is filled with alkali–silica gel. The cement matrix adjacent to the gel-filled crack appears portlandite depleted in XPT; ×35.

**226** Concrete with both DEF and ASR. Tuff coarse aggregate particle (lower left) exhibits both gel-filled internal cracks indicative of ASR and an ettringite-filled peripheral parting indicative of DEF. Cracks running through the cement matrix are mainly filled with ettringite. The cement matrix is shown dark brown; PPT, ×35.

However, it is important to remember that the relative intensity of evidence for each mechanism does not necessarily correspond with the importance of the process as a causal factor in any damage or deterioration. In diagnosis, therefore, petrographers must analyse the observations as well as simply recording the descriptive details (Sims *et al.*, 2004).

## FIRE-DAMAGED CONCRETE

Concrete structures most likely to be subjected to fire include buildings such as offices, warehouses, and schools. Other common scenarios involve vehicle fires in car packs or concrete-lined tunnels. Fortunately, even after a severe fire, concrete structures are generally capable of being repaired rather than demolished. Guidance for the assessment, design, and repair of fire-damaged concrete structures is provided in Concrete Society Technical Report No. 68 (Concrete Society, 2008).

The strength of concrete after cooling varies depending on temperature attained, the heating rate, mix proportions, applied loading, and any external sealing that may influence moisture loss from the surface. For temperatures up to 300°C the residual strength of structural quality concrete is not severely reduced. Generally, between 300°C and 500°C the compressive strength reduces rapidly and concrete that has been heated in excess of 600°C will be of no use structurally. 300°C is normally taken to be the critical temperature above which concrete is deemed to have been significantly damaged. Normally concrete exposed to temperatures above 300°C is replaced if possible. Otherwise the dimensions are increased (for example, by reinforcing columns), depending upon the design loads.

Visually apparent damage induced by heating includes spalling, cracking, surface crazing, deflection, colour changes, and smoke damage. Visual survey of reinforced concrete structure can be performed using a classification scheme from Concrete Society Technical Report No. 68 (2008). This system uses visual indications of the degree of damage to assign each structural member a class of damage from 1 to 5. Each damage classification number has a corresponding category of repair, ranging from decoration to major repair. The Concrete Society classification system is summarized in *Table 26*.

Spalling of the surface layers is a common effect of fires and may be grouped into two main types. Explosive spalling is erratic and generally occurs in the first 30 minutes of the fire. A slower spalling (referred to as 'sloughing off') occurs as cracks form parallel to the fire-affected surfaces, leading to a gradual separation of

**Table 26** Simplified visual concrete fire damage classification (after Concrete Society TR68; 2008)

| Class of damage | Features observed | | | | | | |
|---|---|---|---|---|---|---|---|
| | Finishes | Colour | Crazing | Spalling | Reinforcement bars | Cracks | Deflection |
| 0 (Decoration required) | Unaffected | Normal | None | None | None exposed | None | None |
| 1 (Superficial repair required) | Some peeling | Normal | Slight | Minor | None exposed | None | None |
| 2 (General repair required) | Sunstantial loss | Pink/red | Moderate | Localized | Up to 25% exposed | None | None |
| 3 (Principal repair required) | Total loss | Pink/red or whitish grey | Extensive | Considerable | Up to 50% exposed | Minor | None |
| 4 (Major repair required) | Destroyed | Whitish grey | Surface lost | Almost total | Up to 50% exposed | Major | Distorted |

concrete layers and detachment of a section of concrete along some plane of weakness, such as a layer of reinforcement. Forms of cracking include those caused by differential thermal expansion that often run perpendicular to the outer surface. Also, differential incompatibility between aggregates and cement paste may cause surface crazing. Thermal shock caused by rapid cooling from fire-fighting water may also cause cracking. Macrocrack and microcrack patterns are readily observed microscopically with the aid of fluorescent resin impregnation (227) and fluoresence microscopy (228).

The colour of concrete can change as a result of heating and this may be used to indicate the maximum temperature attained and the equivalent fire duration. In many cases, at above 300°C a red discolouration is important as it coincides approximately with the onset of significant strength loss. Any pink/red discoloured concrete should be regarded as being potentially weakened. Actual concrete colours observed depend on the types of aggregate present in the concrete. Colour changes are most pronounced for siliceous aggregates (especially flint/chert) and less so for limestone, granite, and sintered PFA (shows very little colour change). The colour changes are most easily observed in hand specimen and in thin sections viewed in plane-polarized light. Figure 229 shows red discolouration of concrete aggregate in thin section. The red colour change is a function of (oxidizable) iron content and it should be noted that as iron content varies, not all aggregates undergo colour changes on heating. Also, due consideration must always be given to the possibility that the pink/red colour may be a natural feature of the aggregate rather than heat-induced.

Some widely used aggregate materials contain naturally red or pink particles. British examples include Ordovician and Permo-triassic sandstones and quartzites, which are often various shades of red and any sand/gravel deposits that include materials derived from these rocks (e.g. Trent river gravels). In addition, the Thames river gravels may occassionally include naturally red-coloured flints. Care must also be taken when white calcined flints are present as these are commonly incorporated in decorative white concrete panels and are also a common ingredient of calcium silicate bricks.

Petrographic examination is invaluable in determining the heating history of concrete as it can determine whether features observed visually are actually caused by heat, rather than some other cause. In addition to colour changes of aggregate, the heating temperature can be cross-checked with changes in the cement matrix and evidence of physical distress, such as cracking and microcracking. A compilation of the changes undergone

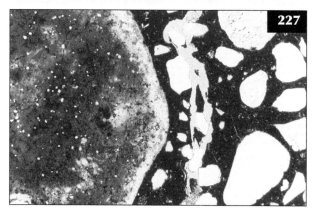

**227** Fire-damaged concrete showing a macrocrack (yellow) running parallel to the outer surface and red discolouration of a flint fine aggregate particle (left). Quartz fine aggregate particles appear white and the cement matrix is dark; PPT, ×35.

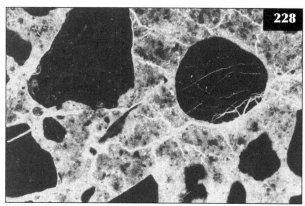

**228** Close view of fire-damaged concrete showing microcracking (green); UV, ×150.

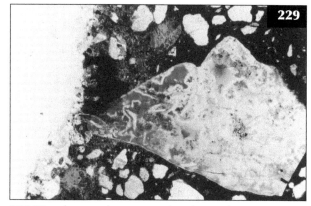

**229** Fire-damaged concrete showing red discolouration of a flint fine aggregate particle near to the spalled outer concrete surface (left), indicating heating to 300–600°C; PPT, ×35.

by concrete as it is heated is presented in *Table 27*. Careful identification of microscopically observed features allows thermal contours (isograds) to be plotted through the depth of individual concrete members. In the most favourable situations contours can be plotted for 105°C (increased porosity of cement matrix), 300°C (red

Table 27 Mineralogical and strength changes to concrete caused by heating (Concrete Society TR68; 2008)

| Heating temperature | Changes caused by heating | |
|---|---|---|
| | Mineralogical changes | Strength changes |
| 70–80°C | Dissociation of ettringite, $Ca_6Al_2(SO_4)_3(OH)_{12} \cdot 26H_2O$, causing its depletion in the cement matrix | Minor loss of strength possible (<10%) |
| 105°C | Loss of physically bound water in aggregate and cement matrix commences causing an increase in the capillary porosity and minor microcracking | |
| 120–163°C | Decomposition of gypsum, $CaSO_4 \cdot 2H_2O$ causing its depletion in the cement matrix | |
| 250–350°C | Pink/red discolouration of aggregate caused by oxidation of iron compounds commences at around 300°C<br>Loss of bound water in cement matrix and associated degradation becomes more prominent | Significant loss of strength commences at 300°C |
| 450–500°C | Dehydroxylation of portlandite, $Ca(OH)_2$ causing its depletion in the cement matrix<br>Red discolouration of aggregate may deepen in colour up to 600°C<br>Flint aggregate calcines at 250–450°C and will eventually (often at higher temperatures) change colour to white/grey<br>Normally isotropic cement matrix exhibits patchy yellow/beige colour in cross-polarized light, often completely birefringent by 500°C | |
| 573°C | Transition of α- to β-quartz, accompanied by an instantaneous increase in volume of quartz of about 5% in a radial cracking pattern around the quartz grains in the aggregate | Concrete not structurally useful after heating in temperatures in excess of 550–600°C |
| 600–800°C | Decarbonation of carbonates; depending on the content of carbonates in the concrete, e.g. if the aggregate used is calcareous, this may cause a considerable contraction of the concrete due to release of carbon dioxide, $CO_2$; the volume contraction will cause severe microcracking of the cement matrix | |
| 800–1200°C | Complete disintegration of calcareous constituents of the aggregate and cement matrix due to both dissociation and extreme thermal stress, causing a whitish grey colouration of the concrete and severe microcracking. Limestone aggregate particles become white | |
| 1200°C | Concrete starts to melt | |
| 1300–1400°C | Concrete melted | |

discolouration of aggregate), 500°C (cement matrix becomes wholly birefringent) (**230**), 600°C (α- to β-quartz transition), 800°C (calcination of limestone), and 1200°C (first signs of melting).

Figure **231** shows some microscopical features that may be observed in fire-damaged concrete (example adapted from Smart, 1999). Some aggregate particles have been reddened indicating that the concrete has reached at least 300°C at that point. Particles of flint have been calcined and so have been heated to 250–450°C. The cement matrix is bisected by numerous fine cracks, some of which radiate from quartz grains in the fine aggregate fraction. This deep cracking and cracking associated with quartz suggest that the concrete has reached 550–575°C. Overall we can deduce that the concrete has been heated to approximately 600°C in the area represented by the sample.

By determining the position of thermal contours through the cross-section of a concrete element, an estimate can also be made of the likely condition of reinforcement bars. At 200–400°C prestressed steel shows considerable loss of strength, at >450°C cold-worked steel loses residual strength, and at >600°C hot-rolled steel loses residual strength.

## HIGH-ALUMINA CEMENT (HAC) CONCRETE

HAC is a type of calcium aluminate cement, which is manufactured by fusing limestone and bauxite, rather than the limestone and clay/shale used for Portland cement. HAC was used in structural concrete mainly from the 1950s to the early 1970s, most commonly in the manufacture of precast, prestressed concrete beams. Although more expensive than Portland cement, HAC was popular in the precast industry for its high early strength that provided higher production rates from moulds.

Unfortunately, in the early 1970s several structural failures occurred and it was determined that HAC undergoes a natural and inevitable reaction known as 'conversion'. Conversion comprises the mineralogical change of the HAC paste from metastable hydrates to their stable form over a period of years/decades. This results in HAC concrete with increased porosity and significantly lower strength. Consequently, by 1976 HAC was effectively banned for use in structural concrete and it remains so today. Calcium aluminate cements are still used for specialist products such as grouts, repair mortars, and refractory products (Concrete Society, 1997).

A considerable amount of the existing building stock contains structural HAC concrete elements, for example

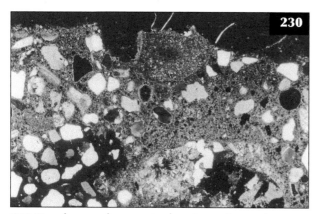

**230** Fire-damaged concrete showing anisotropic properties and yellow–beige colour of the cement matrix, indicating heating to 500°C; XPT, ×35.

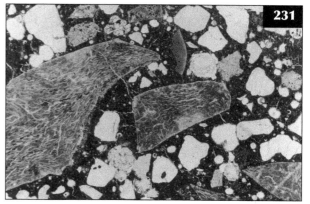

**231** Fire-damaged concrete showing both reddened (red) and calcined (brown mottled) flint aggregate particles. There are numerous fine cracks (white) within the cement matrix (dark), some of which radiate from quartz fine aggregate (white); PPT, ×35.

50,000 buldings in the United Kingdom alone (Rushton, 2006). These must all be subject to investigations and monitoring to ensure satisfactory performance. In general terms, HAC concrete investigations comprise identification of members containing HAC, strength appraisal, and durability assessment. This has created a demand for petrographic examination of lump samples taken from HAC concrete beams (typically sampling 5% of all beams present) to provide definitive identification of cement type and assess a number of parameters affecting durability.

Optical microscopy is the most popular definitive method of identifying the presence of HAC in concrete. Unhydrated HAC clinker consists mainly of calcium aluminates with much smaller amounts of ferrite and silicate phases. Once hydrated, HAC consists largely of calcium aluminate hydrates. In cross-polarized light in thin section, the appearance of hydrated HAC is largely isotropic, if uncarbonated, and light brown birefringence, if carbonated. 232 shows a comparison of the general microscopical appearance of HAC concrete and Portland cement concrete from a structure where one was cast directly against the other. In plane-polarized light many residual grains of unhydrated and partially hydrated clinker are usually seen, which are often opaque or red (233). These phases cannot be identified with certainty in thin section, although ferrites are usually assumed to be one of the constituents (Hewlett, 1998). One characteristic constituent that can be identified is 'pleochroite', which is fibrous and has distinctive blue to white pleochroism (234).

In hand specimen, the HAC used in British structural concrete was usually dark greyish-brown in colour. Although colour may suggest that HAC is present it should not be used as a basis for identification as certain Portland-type cements may also exhibit dark colours. Calcium

**232** Precast HAC concrete (left) with a carbonated HAC matrix (brown) and *in situ* Portland cement (right) with an uncarbonated Portland cement matrix (black); XPT, ×35.

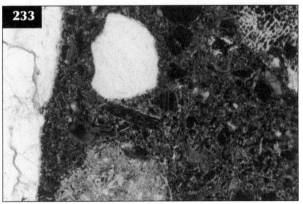

**233** Close view of HAC concrete illustrating the appearance of the cement matrix; PPT, ×150.

aluminate cements presently manufactured exhibit colours ranging from white, grey, buff or black (Hewlett, 1998).

The strength of HAC concrete is normally determined directly by compressive strength testing of core samples. As all of structural HAC concrete beams in the UK are more than 30 years old, they are now assumed to be fully converted and to have reached their stable minimum strength. Therefore, assessment of the degree of conversion is rarely required; if necessary this is best estimated by differential thermal analysis (DTA).

The main durability issues affecting HAC concrete beams are reinforcement corrosion and chemical attack (Dunster, 2002b). Typically these would be investigated by optical microscopical examination in thin section. The likelihood of carbonation-induced reinforcement corrosion is determined by measuring the depth of carbonation microscopically (and determining whether carbonation has reached the prestresssing wires of the beam). The carbonation reaction of HAC is different to that of Portland cement, with calcium aluminate hydrates reacting with carbon dioxide from the atmosphere to form a mixture of calcium carbonate (chiefly calcite) and aluminium hydroxide (chiefly gibbsite). Figure 235 illustrates the contrasting appearance of carbonated and uncarbonated HAC matrix. For the pre-1976 HAC concrete found in the UK, the shallow depth of cover concrete (typically 20 mm) is usually fully carbonated with alkaline protection of the prestressing wires being lost.

Where HAC concrete has been exposed to prolonged wet conditions it may be subjected to deleterious chemical attack. The two forms of chemical attack applicable to HAC concrete are sulfate attack and alkaline hydrolysis. These are detected by a combination of deterioration seen with optical microscopy and high

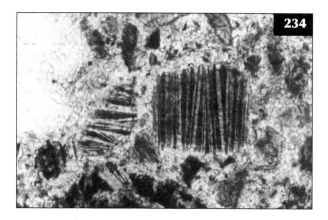

**234** Very close view of HAC concrete showing grains of pleochorite (centre right, blue/black striped and centre left, white/black striped); PPT, ×600.

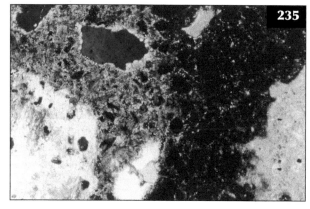

**235** HAC concrete with the left half appearing carbonated (light brown cement matrix) and the right half appearing uncarbonated (dark brown cement matrix); XPT, ×150.

concentrations of sulfate or alkalis determined by chemical analysis. Sulfate attack occurs when HAC concrete comes into contact with solutions leached from sulfate-bearing materials (e.g. gypsum plaster). In thin section, sulfate attack is characterized by the presence of ettringite in cracks and voids, and degradation of the cement matrix. Alkaline hydrolysis occurs when persistent water leakage causes sodium and potassium hydroxide solutions to leach onto HAC concrete. In thin section, this may be indicated by the presence of leached secondary deposits of calcium carbonate and cement matrix replacement (236) and/or cracking (237).

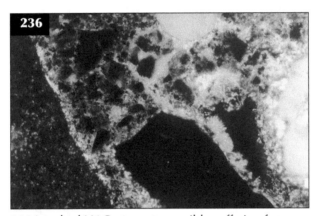

**236** Leached HAC concrete possibly suffering from alkaline hydrolysis. Secondary deposits of calcium carbonate (pink) are seen replacing the cement and filling air voids (right); XPT, ×35.

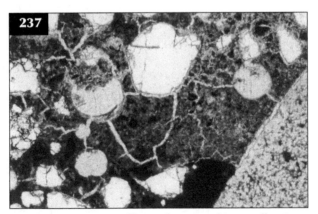

**237** HAC concrete suffering from cracking (yellow) induced by alkaline hydrolysis; PPT, ×150.

# Concrete products

**6**

## INTRODUCTION

Concrete products are concrete construction units or modules that are built into structures after they have fully hardened. They cover a wide range of products that are often 'precast' in that they are cast in moulds, usually in a factory, providing economies of scale and improved quality control. Concrete products form a significant portion of the construction materials industry and may be divided into the following groups:

- Architectural cast stone and other types of ornamental concrete.
- Load-bearing structural units such as piles, lintels, beams, railway sleepers, pipes, and water tanks.
- Concrete steps, paving flags, block pavers, and kerbs.
- Concrete blocks and slabs.
- Roofing tiles.

Architectural cast stone is used as a substitute for natural stone on buildings where cost savings are required. It comprises concrete units made from ingredients that are carefully chosen to simulate the appearance of natural stone. It is rarely seen on buildings dated before 1900 but is now widely used for external masonry in areas with planning restrictions, which require a stone-like appearance.

Precast concrete products have been in use since the mid-19th century and they now comprise a major proportion of the concrete industry. They include various load-bearing units which are used either alone to form concrete structures, or in combination with *in situ* concrete. Many of these units are prestressed and in some cases the appearance is of secondary importance. Other precast units are not load-bearing and these include items which are required to resist wear by foot and other traffic by having a strong dense wearing surface. The manufacture of load-bearing and wear-resistant units is often conducted in accordance with normal concrete practice. In contrast, precast concrete units for masonry wall construction are usually turned out from machines in which a semi-dry mix is heavily rammed or pressed into a compressible steel

mould. Lightweight aggregate blocks, manufactured using clinker aggregate, are used extensively for internal masonry building partitions. Another form of lightweight concrete is 'aircrete' which is an aerated cement mortar. This has been used for semistructural masonry blocks and floor/roof panels since the 1950s.

Calcium silicate units differ from the other concrete products, as they do not contain Portland cement. Instead, they are manufactured from siliceous aggregate and lime; there is a range of products, the most important being masonry bricks. Invented and patented in Britain in 1866, calcium silicate bricks are now used extensively throughout Europe and are the dominant brick type in Holland and Germany (Bowley, 1993/4).

Fibre reinforcement is used in some concrete products to enable the production of building elements of reduced minimum dimensions and increased flexural and impact strengths. Asbestos was first used in widely manufactured concrete products in the late 19th century and, since then, other fibres made from glass, steel, polymers, carbon, and vegetable matter have all been utilized.

The principal uses of petrographic examination for investigations of concrete products are:

- Quality assessment during product development and manufacture.
- Identifying the type of concrete product in existing structures.
- Identifying the ingredients used and the manufacturing process.
- Screening the product for inherent defects including potentially deleterious ingredients.
- Identifying the presence of hazardous materials such as asbestos.
- Diagnosing the causes and significance of decay and deterioration in service.

For most concrete products, the petrographic techniques and examples presented in Chapter 5 (Concrete) will apply and these should be referred to. To avoid repetition, four of the more specialized types of concrete product have been chosen for detailed discussion in the following pages.

These have been chosen to illustrate how petrography can be applied to concrete products, rather than being an exhaustive account of all product types.

## PETROGRAPHIC EXAMINATION

Petrographic examination of concrete products is performed in accordance with ASTM C856 (ASTM International, 2004). Once a sample is received in the laboratory, an initial visual and low-power microscopical examination is conducted to determine the number of

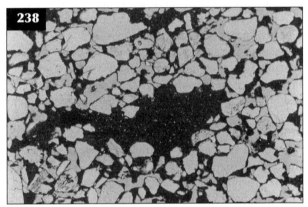

**238** Cast stone manufactured by a 'semi-dry' process, showing quartz fine aggregate particles (white), white Portland cement matrix (dark brown), and air voids (yellow). An unmixed lump of cement is seen in the centre of view; PPT, ×35.

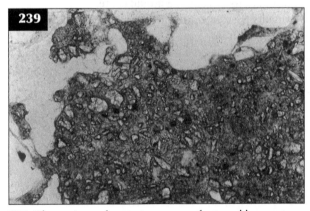

**239** Close view of cast stone manufactured by a 'semi-dry' process, showing white Portland cement matrix (brown) including an abundance of unhydrated white Portland cement grains. Quartz fine aggregate particles appear white and air voids are shown yellow; PPT, ×300.

material types, dimensions, manufacturing marks, colour, and the relative hardness and coherence of the material. The study of defects such as cracking may be enhanced by additional low-power microscopical examination of a finely polished slice. High-power examination of thin section specimens is then used to determine the grading and mineralogy of the aggregate, cement type, the presence of mineral additions/pigments, assess the quality of workmanship, and screen the material for evidence of distress or deterioration.

## ARCHITECTURAL CAST STONE

Cast stone units consist of concrete intended to resemble, and be used in a similar way to, natural stone. The desired colours and textures are obtained by blending white and ordinary Portland cement with inert pigments, and by the use of aggregates of appropriate colour and texture. The aggregate used is often crushed rock derived from the same natural stone that the cast stone is intended to imitate. Units may be homogeneous throughout or consist of a facing mix and a backing mix. It is common practice only to have a 20–30 mm thickness of the expensive decorative facing mix on a backing mix of cheaper normal concrete.

The units are manufactured either by a 'semi-dry' process (also known as 'vibrant dry tamp') or by a 'wet cast' process. Semi-dry units normally comprise a facing mix and a backing mix. However, if the shape of the element is complex or it is less than 75 mm thick, the facing mix is used throughout. The semi-dry process involves the use of a low water content ('earth moist') mix that has relatively poor workability. It has to be mixed using a forced action mixer and compacted into moulds using pneumatic or electric rammers. The benefit is that the casting can be immediately removed from the mould, giving high productivity from each mould. Other advantages include low drying shrinkage and a relatively open textured finish that resembles certain types of natural stone and often requires no surface treatment to achieve the desired finish. However, the porosity of the open texture can be a disadvantage in terms of durability and chemical admixtures (usually metallic stearates) are commonly incorporated to improve the waterproofing (United Kingdom Cast Stone Association, 2004).

In thin section, semi-dry cast stone typically exhibits an abundance of small, interconnected entrapped air voids (up to 30% excess voidage) and may exhibit some 'balling' of cement caused by workability difficulties. Figures **238** and **239** show a semi-dry cast stone facing mix made from fine quartz sand and white Portland cement that exhibits a characteristically high entrapped air void content and minor balling of the cement. The overall visual appearance in hand specimen is reminiscent of sandstone.

The wet cast process uses similar mix ingredients to the semi-dry process but with higher water content, giving good workability. Wet cast units are usually homogeneous with one mix being used throughout. The mix is easily compacted using conventional concreting techniques, such as poker or table vibration. The disadvantages are lower production rates (as each mould can only be used once per day) and that the relatively wet mix is more prone to shrinkage cracking and surface crazing. In addition, significant work is required to surface finish each unit because of the formation of cement laitance against the mould face. The thin laitance layer can be removed by acid etching, dry rubbing, sand blasting, or wet grinding to reveal a uniform decorative finish.

In thin section, wet cast stone appears similar to concrete except for the use of decorative ingredients. Figures **240** and **241** show a wet cast unit comprising brown limestone, bound by white Portland cement and exhibiting a low excess voidage (<3%). The cement matrix is seen to exhibit microcracking caused by drying shrinkage (**242**). The overall visual appearance of the sample in hand specimen is similar to compact light brown limestone.

Many cast stone units, especially slender ones such as sills, mullions, and copings, incorporate steel reinforcement bars. This increases their tensile strength and is primarily intended to help units withstand the stresses of handling. Reinforcement should ideally be embedded in wet cast units or the backing mix of semi-dry units in order to prevent corrosion of the steel.

As cast stone is in fact concrete, it will weather and deteriorate as concrete and can exhibit a similar range of defects. The most common defects seen in cast stone are mechanical damage during handling, surface crazing, and reinforcement corrosion. Damage during handling is usually seen as large cracks, impact marks, or lost corners that are first visible prior to installation of the unit.

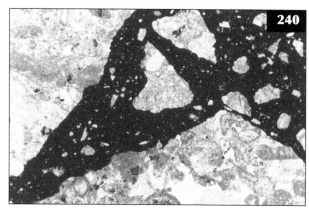

**240** General view of 'wet cast' cast stone showing decorative limestone aggregate (light brown/pink) bound by uncarbonated white Portland cement matrix (dark brown); XPT, ×35.

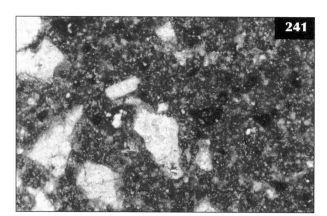

**241** Close view of 'wet cast' cast stone showing limestone aggregate (light brown) bound by uncarbonated white Portland cement matrix (dark brown). Unhydrated/partially hydrated cement grains are sporadically seen (right of centre); XPT, ×150.

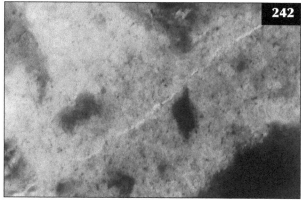

**242** Close view of 'wet cast' cast stone showing a microcrack (yellow) caused by drying shrinkage. The fluorescence of the cement matrix (green) indicates moderate microporosity and W/C in the normal range; UV, ×150.

Surface crazing is a pattern of fine cracking seen on the outer surface that is caused by drying shrinkage of a wet mix or imperfect curing techniques. Surface crazing is liable to become more visible with time as the fine cracks fill with dark dirt that contrasts with the typically light colour of the unit. Crazed units will carbonate faster and increase the risk of corrosion damage in reinforced units. In unreinforced units, crazing is less of a durability issue but may cause complaints for aesthetic reasons. By embedding steel reinforcement into cast stone the manufacturer builds in a latent defect that would not be present in natural stone. Carbonation-induced corrosion of embedded steel reinforcement is a common cause of cracking and spalling of cast stone during service. Reinforcement placed within porous facing mix layers, or at the interface between facing and backing mix layers is particularly susceptible to corrosion.

## AIRCRETE PRODUCTS

Aircrete, also known as autoclaved aerated concrete (AAC), is commonly used for the manufacture of mass-produced lightweight concrete blocks. In addition, it is produced as panels with steel reinforcement added, to improve flexural and shear capacity (reinforced autoclaved aerated concrete, RAAC). Aircrete is an aggregate-free, aerated mortar with low density and good insulating properties. It is manufactured by foaming a slurry of Portland cement (or lime) mixed with siliceous filler, such as PFA, GGBS, or silica flour. Chemical admixtures and inert pigments are frequently added. The aeration is commonly achieved chemically by adding aluminium powder to the wet slurry. This reacts with the alkaline lime (calcium hydroxide) within the cement to produce bubbles of hydrogen gas as shown in the following equation:

$$CaOH_{2(s)} + 2Al_{(s)} + 2H_2O_{(l)} \rightarrow CaAl_2O_{4(s)} + 3H_{2(g)}$$

Alternatively, aeration may be achieved by mechanical means such as entraining air by a whipping process, or by adding preformed stable foam to the slurry. The slurry is poured into a steel mould where it sets to form a weak 'cake', which is cured for several hours at elevated temperature before demoulding, trimming, and final curing in an autoclave.

In thin section, the texture of aircrete is dominated by air voids, which may comprise as much as 80% of the total volume (Building Research Establishment, 2002b). Air voids are typically spherical and less than 2 mm in diameter. The materials of the hardened slurry matrix are finely divided and abundant remnants of unhydrated cement grains and filler should be visible at greater than ×100 magnification. Owing to the porous nature of aircrete, the matrix would be expected to carbonate fairly rapidly during service. Figures 243–245 show an aircrete masonry block comprising aerated Portland cement with a filler of PFA.

The petrographer may be called upon to investigate the following problems that are sometimes associated with aircrete products. Aircrete masonry blocks may exhibit excessive swelling and shrinkage (unsoundness) if unstable ingredients are included in the mix. Of particular concern is PFA filler that can potentially contain excessively high contents of unburnt coal (BS EN 450 specifies a loss on ignition of ≤7%).

Concerns have been raised regarding the in-service performance of RAAC panels designed before 1980. There have been many instances where they have exhibited excessive deflection (sagging), with associated tension cracking sometimes visible on the soffit face (Building Research Establishment, 2002b). Excessive voidage around the reinforcement and/or reinforcement corrosion may contribute to deflection. Reinforcement used in RAAC panels must have a protective coating as the porous aircrete provides insufficient alkaline protection. If the protective coating is inadequate or fails, the unit is vulnerable to reinforcement corrosion, especially in damp environments.

## CALCIUM SILICATE UNITS

Calcium silicate units are manufactured from fine siliceous aggregate (and sometimes silica flour) and 10–20% ground quicklime or well hydrated lime. Inert pigments may be included to influence the final brick colour. The units are shaped into moulds under pressure and cured by autoclaving at around 170°C for between 4 and 16 hours. During curing some of the lime reacts with the siliceous aggregate to form calcium silicate hydrate, which binds the brick together.

The term 'sandlime' is used to describe calcium silicate bricks in which only natural quartz sand comprises the siliceous portion of the raw ingredients. 'Flintlime' bricks include a substantial proportion of crushed flint (Building Research Establishment, 1992).

In thin section, the appearance of calcium silicate brick is dominated by the presence of unreacted siliceous aggregate particles. New bricks exhibit a binder of calcium silicate hydrates that appear isotropic. Old bricks tend to be partially or fully carbonated with the binder including, or composed of, finely crystalline calcium carbonate. Calcium silicate bricks typically appear fully compacted with no large entrapped air voids present. The appearance of a sandlime brick is illustrated in Figures 246–248, and

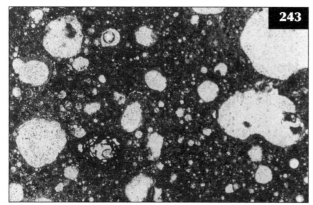

**243** General view of aircrete masonry block in thin section. Cement matrix appears dark brown and air voids are shown yellow; PPT, ×35.

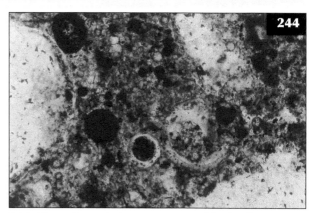

**244** Close view of aircrete masonry block. Showing Portland cement matrix including abundant cenospheres of PFA (black/white). Air voids are shown yellow; PPT, ×300.

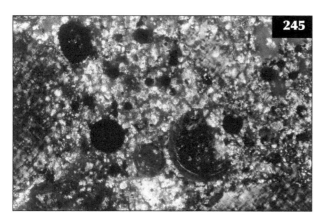

**245** Same view as **244** in cross-polarized light, showing carbonated cement matrix (brown) including isotropic cenospheres of PFA (black). Air voids are shown dark green; XPT, ×300.

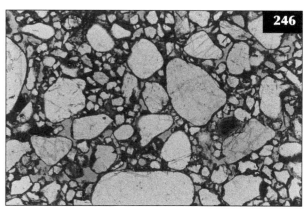

**246** General view of sandlime brick showing quartz fine aggregate particles (white), lime/calcium silicate hydrate binder (brown), and small air voids (yellow); PPT, ×35.

**247** Same view as **246** in cross-polarized light showing quartz fine aggregate particles (grey/white) and carbonated lime/calcium silicate hydrate binder (brown); XPT, ×35.

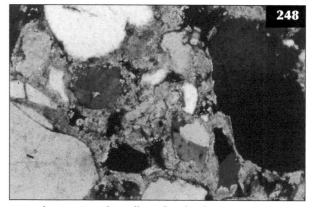

**248** Close view of sandlime brick, showing quartz fine aggregate particles (grey/white) bound by a matrix of carbonated lime/calcium silicate hydrate (light brown); XPT, ×150.

a flintlime brick is illustrated in Figures 249–251. Both of these examples are old calcium silicate bricks that are fully carbonated.

Calcium silicate brickwork is more prone to expansion and shrinkage than clay brick masonry and in certain circumstances this may cause significant masonry cracking. This has caused some to regard calcium silicate bricks as potentially deleterious and, consequently, there is often a requirement to identify their presence during building surveys. This identification is routinely done by petrographic examination of small lump samples, which is regarded as the definitive method.

## ASBESTOS CEMENT PRODUCTS

Asbestos cement products are the most common asbestos-containing materials in use and their manufacture consumes 70% of world asbestos production. The range of products includes wall and roof sheet (corrugated and flat), roof tiles, rainwater goods, sewerage pipe, and pressure pipe. Asbestos cement products are manufactured using mechanized processes with 10–20% asbestos fibre being mixed with Portland cement slurry. After forming into the desired shape they are usually steam cured with low pressure steam at 80°C,

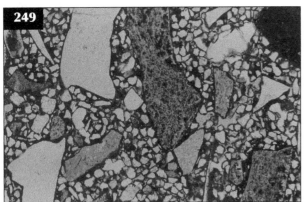

**249** General view of flintlime brick showing crushed flint (white/brown), quartz (white) fine aggregate (white), lime/calcium silicate hydrate binder (brown). Some of the flint aggregate particles appear to have been calcined prior to brick manufacture (mottled brown); PPT, ×35.

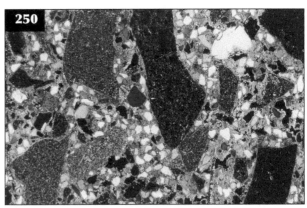

**250** Same view as **249** in cross-polarized light showing crushed flint (grey), quartz (white) fine aggregate particles, and carbonated lime/calcium silicate hydrate binder (light brown); XPT, ×35.

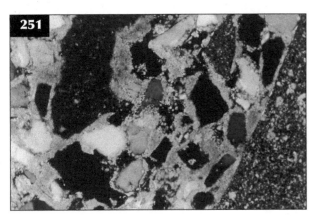

**251** Close view of flintlime brick showing flint (grey) and quartz (black/grey/white) fine aggregate particles bound by a matrix of carbonated lime/calcium silicate hydrate (light brown); XPT, ×150.

or high pressure steam at 170°C. Up to 40% of the cement may be replaced by silica flour, if high pressure steam is to be used for curing. Curing is sometimes achieved at ambient temperature or by autoclaving.

In thin section, asbestos cement typically consists of finely divided asbestos fibre dispersed in a matrix of Portland cement, including abundant partially hydrated relict cement grains. Asbestos cement is usually manufactured using chrysotile (white asbestos), although crocidolite (blue asbestos) and amosite (brown asbestos) are used for some products. The asbestos fibres and fibre bundles usually exhibit a preferred orientation subparallel to the length of the sheet, tile, or pipe that they are incorporated within. Traces of the asbestos host rock such as serpentinite, olivine, or pyroxene may also be present. If silica flour was included, remnants may be visible in the cement matrix. Asbestos cement products would be expected to appear uncarbonated when new and over time the cement matrix will carbonate. Figures 252 and 253 show the appearance of a fully carbonated, flat asbestos cement sheet in thin section.

Although asbestos cement products are usually very durable in service, the petrographer may be called upon to investigate a number of issues. The most common are confirming that hazardous asbestos is present and determining the composition and manufacturing process of the products. Deterioration issues that can be studied include carbonation leading to embrittlement (and associated loss of strength) and attack by aggressive water on pipes, due to soft water or sulfates.

Owing to the hazardous nature of asbestos, all sampling and handing of products suspected to contain asbestos should be performed using appropriate safe working procedures (Health and Safety Executive, 2005). Also, the petrographer should be aware of any legal obligations regarding asbestos. For example, in the United Kingdom the Control of Asbestos at Work Regulations apply to all work activities involving asbestos-containing materials.

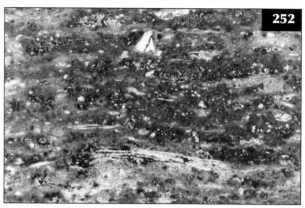

**252** General view of asbestos cement sheet consisting mainly of carbonated Portland cement (brown) with abundant unhydrated/partially hydrated cement grains and asbestos fibres (bright, elongated); XPT, ×35.

**253** Close view of a chrysotile (white asbestos) fibre bundle within asbestos cement sheet. The chrysotile exhibits first-order yellow interference colours. The fibre is surrounded by carbonated Portland cement (brown) that incorporates unhydrated/partially hydrated cement grains; XPT, ×150.

# Floor finishes

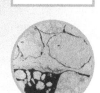

## INTRODUCTION

A wide variety of finishing materials are applied to internal floor surfaces for one or more of three reasons:

- To protect the structural elements of the floor from wear and deterioration.
- To provide an attractive appearance for the floor.
- To increase comfort and safety for the floor user.

Floor coverings that have been used since antiquity include geomaterials such as stone, brick, and cementitious materials, as well as other types of material including timber and textiles. In the last 100 years the number of floor finishes available has increased greatly and in recent years these include materials based on, or

incorporating, polymers. A selection of flooring materials that may be encountered by the petrographer is categorized in *Table 28*.

Several of these materials used for flooring are adequately covered in other chapters (i.e. stone, brick, ceramic tile, concrete, and asphalt) and certain other materials are not within the scope of this book. This chapter will concentrate on some of the remaining flooring materials that are most commonly examined by the petrographer, namely screed, terrazzo, and synthetic resin.

Petrographic examination of samples taken through the various layers of a flooring system can provide considerable information on the nature and relationship of the flooring materials (including the substrate) and their interfaces.

**Table 28** Flooring materials that may be encountered by the petrographer

| Jointless floor finishes | Jointed resilient floor finishes | Jointed hard floor finishes |
| --- | --- | --- |
| Concrete wearing surfaces | Textile | Ceramic tiles |
| Polymer-modified cementitious screeds | Linoleum | Brick pavers |
| Granolithic and cementitious wearing screeds | Cork | Natural stone |
| *In situ* terrazzo | Timber products | Terrazzo tiles |
| Synthetic resins | Polyvinyl chloride (PVC) | Composition block |
| Mastic asphalt | Rubber | Metal |
| Magnesium oxychloride | Thermoplastic tiles | |

The principal applications are:

- Identifying the type of flooring products used, their ingredients, and the manufacturing process.
- Assessing workmanship issues such as compliance with the specification for installation procedures.
- Diagnosing the causes and significance of perceived flooring defects and failures during service.
- Investigating surface staining, discolouration, and other aesthetic issues.

## PETROGRAPHIC EXAMINATION

Petrographic examination of cementitious flooring materials can be performed in accordance with ASTM C856 (ASTM International, 2004). Noncementitious materials can also be examined by adapting the ASTM method. Once a sample is received in the laboratory, an initial visual and low-power microscopical examination is conducted to determine the number of layers, their thickness, colour, relative hardness, relative bond strength, and coherence. This examination can often be usefully supplemented by examination of a finely ground slice cut through the various layers. High-power examination of thin section specimens is then used to identify the material types and their ingredients, assess the quality of workmanship, and screen the materials for evidence of distress or deterioration.

## FLOOR SCREED

Screeds are laid either over the structural floor to provide a suitable level surface for the floor finish, or as the floor wearing surface.

Screeds which are used to provide a smooth and level base for subsequent laying of floor finishes include sand:cement screed, lightweight screeds, and calcium sulfate screeds. Most commonly used for this purpose are sand:cement screeds, which may be designed to be fully bonded to the structural base, unbonded, or floating (laid immediately over a compressible layer such as expanded polystyrene). Sand:cement screeds consist of fine aggregate and Portland cement with mix proportions ranging from 1:3 to 1:4 (cement:sand, by weight), depending on the anticipated degree of traffic. For screeds of over 50 mm thick, some 10 mm coarse aggregate may be incorporated into the mix to control drying shrinkage. In order to achieve the required tolerance in levels, sand:cement screeds are made using the minimum amount of water and for this reason they are termed 'semi-dry'. This low water content inhibits screed workability and consequently pan type (forced action) mixers must be used to ensure that the cement is adequately distributed. Figure **254** shows the typical appearance of semi-dry sand:cement screed consisting of natural sand fine aggregate and Portland cement. Polypropylene fibres have been added to this screed to help reduce drying shrinkage. Semi-dry sand:cement screeds are usually open textured with a common abundance of entrapped air voids (say up to 15%), (**255**). These high air void contents may result from one of, or a combination of two causes. Firstly, the screed may have been under-compacted and, secondly, the screed may have started to set before it was laid, with the associated stiffening of the cement paste reducing workability. Also, if screeds are laid as floating screeds on a layer of insulation, the compressible nature of the insulation may hinder compaction to some degree. Historically, semi-dry sand:cement screeds have been the source of numerous flooring failures (Gatfield, 1998). Unsatisfactory low strengths of sand:cement screed may result from an excessively high air void content (up to 40% is known), a lower than adequate cement content or uneven distribution ('balling') of the cement. Figure **256** shows a very poorly compacted semi-dry type sand:cement screed that was unsatisfactorily weak and friable and collapsed in service. Other defects of sand:cement screeds that may require investigation include cracking, curling, and debonding, associated with inadequate curing and drying shrinkage.

Lightweight screeds are less robust than sand:cement screeds and are used to save weight or provide insulation. They are Portland cement-based and are made lightweight either by the use of lightweight aggregate (e.g. sintered PFA) or by aeration.

Calcium sulfate screeds are based on either anhydrite or hemihydrate. They have the advantages that they can be laid much thinner and set faster than sand:cement screeds and the disadvantage that they must be protected from moisture in service. Figures **257-259** show the appearance of anhydrite screed in thin section, consisting of natural sand fine aggregate bound by a matrix of gypsum crystals. Calcium sulfate screeds must be kept dry at all times since they rapidly lose strength in the presence of water. Portland cement-based materials laid in contact with calcium sulfate screeds are potentially susceptible to sulfate attack in the presence of moisture.

Screeds suitable for use as wearing surfaces include polymer-modified cementitious screeds and granolithic (and other cementitious) wearing screeds. Polymer-modified cementitious screeds are commonly used where a thin wearing screed layer is to be laid on a concrete base. Types of polymers added to the Portland cement binder include styrene butadiene rubber (SBR), acrylics, and bitumen. A wide range of aggregates and fillers may be incorporated including cork, rubber, hardwood, sand, slate, and wood flour (Pye & Harrison, 1997).

**254** 'Semi-dry' sand:cement screed, showing quartz-rich natural sand fine aggregate (mainly quartz, grey/white) bound by a matrix of uncarbonated Portland cement (black). A monofilament polypropylene fibre runs through the field of view (brightly coloured). Air voids appear dark green; XPT, ×35.

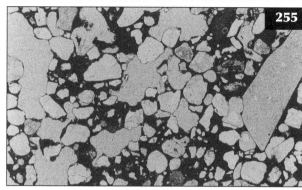

**255** 'Semi-dry' sand:cement screed, showing a cement-rich lump that is the result of inadequate mixing (lower right) and the remainder of the view exhibits abundant entrapped air voids (yellow) caused by poor compaction. Aggregate particles appear white, the Portland cement matrix appears dark brown; PPT, ×35.

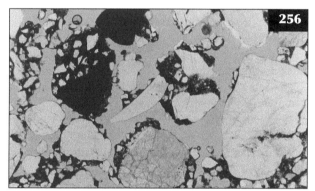

**256** Very poorly compacted 'semi-dry' sand:cement screed showing a very high abundance (40%) of entrapped air voids (yellow). Natural sand fine aggregate particles appear white/black and Portland cement binder is shown dark brown; PPT, ×35.

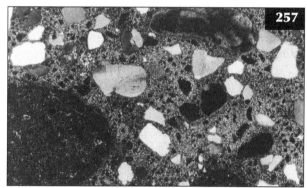

**257** Anhydrite screed comprising natural sand fine aggregate including particles of quartz (white, grey), flint (brown/grey), and glauconite (green), bound by a hardened matrix of gypsum (grey); XPT, ×35.

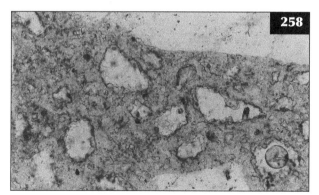

**258** Close view of anhydrite screed showing the matrix of gypsum crystals (light brown). Fine aggregate particles appear white and air voids are shown yellow; PPT, ×300.

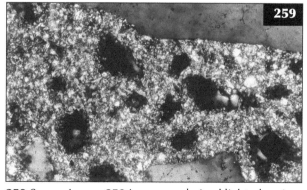

**259** Same view as **258** in cross-polarized light, showing the matrix of gypsum crystals (light grey) of an anhydrite screed. Quartz fine aggregate particles appear grey and air voids are shown dark green; XPT, ×300.

Granolithic screed is essentially concrete which contains granite or other suitable crushed rock aggregate, to provide wear resistance. Figure **260** shows the 'concrete-like' appearance of a granolithic screed consisting of granite coarse aggregate and natural sand fine aggregate, bound by a Portland cement matrix; the screed exhibits typical good compaction with very few air voids present. Granolithic screeds usually provide durable floor finishes but problems do occasionally occur. These typically involve drying shrinkage cracking as granolithic screeds are normally cement-rich and can have a high water demand. Other defects may involve debonding and/or curling of the screed, due to drying shrinkage and possibly inadequate preparation of the base.

## TERRAZZO

Terrazzo is a special form of decorative concrete floor finish that uses coloured cement (often white) and decorative aggregate such as marble chips. The surface is ground to expose the aggregate, which has a great influence on the appearance.

Terrazzo may be laid *in situ* or, more commonly, as terrazzo tiles. *In situ* terrazzo is normally laid on a high-quality screed while the screed is still 'green', to a thickness of not less than 15 mm. Once laid, coarse surface grinding is undertaken after about 4 days, then filling and fine grinding after a further 3 days. Terrazzo tiles consist of similar mixes to *in situ* terrazzo, but they are precast in factories under controlled conditions and are now generally preferred owing to their assured quality. Tile thickness varies depending on the plan size with a common tile size being 300 × 300 × 28 mm. The tile thickness may include a nondecorative (and cheaper) backing mix.

In thin section, terrazzo typically has the appearance of fine aggregate-deficient and cement-rich concrete (cement:aggregate ratio typically 1:2.5–3.0). The cement type is often white Portland cement and may include a considerable quantity of pigment. Owing to the high cement content it is relatively common to observe some degree of microcracking of the cement matrix. Figure **261** shows the appearance of a cast *in situ* terrazzo consisting of marble chips bound by white Portland cement. The fine detail of the uncarbonated cement matrix is obscured by the presence of pigment. Figure **262** shows a terrazzo floor tile consisting of polishable crushed limestone coarse aggregate and crushed decorative marble fine aggregate, bound by a white Portland cement matrix. Again, the detail of the uncarbonated cement matrix is obscured by the presence of pigment and also a certain amount of calcium carbonate dust.

Providing that it is correctly formulated, laid, and maintained, terrazzo can provide a relatively hard and durable floor surfacing. Some of the most common terrazzo defects involve unsightly macrocracking that runs perpendicular to the upper surface. *In situ* terrazzo is

**260** Granolithic screed with crushed granite coarse aggregate (upper left) and natural sand fine aggregate, bound by an uncarbonated matrix of Portland-type cement (black with brightly coloured 'specks' comprising portlandite crystallites). Well compacted with no air voids visible; XPT, ×35.

**261** *In situ* terrazzo floor finish, comprising crushed marble chips (pink/light brown) bound by an uncarbonated matrix of pigmented, white Portland cement (dark brown); XPT, ×35.

susceptible to shrinkage macrocracking and, consequently, should be laid in bays of no more than 1 m² in size, subdivided by strips of jointing material. Avoidance of shrinkage cracking is one reason that there has been much greater use of terrazzo tiles (Pye & Harrison, 1997). Other causes of macrocracking include structural movement, reflective cracking, shrinkage of bedding, weak bedding, and use of tiles that are too thin for the imposed load.

The finely ground surface of terrazzo may become polished and slippery with wear, especially when wet. Terrazzo can be particularly slippery when polished with wax or other polishing compounds. Inappropriate cleaning techniques can cause surface pitting and the use of certain cleaning chemicals can potentially cause staining and discolouration of the upper surface. With time terrazzo floors can become unsightly with scratched surfaces that are ingrained with dirt. Figure **263** shows the upper surface of a terrazzo floor finish viewed in oblique reflected light with clearly visible scratches caused by wear in service. Superficial terrazzo surface defects can be removed and slip resistance maintained by light regrinding.

## SYNTHETIC RESIN FLOOR COVERING

Resin floorings were developed in the 1960s and are available in a wide range of products from thin floor seals to heavy duty industrial protection. They provide jointless floor covering that has proved most popular in special industrial buildings used for food provisioning, pharmaceuticals, and electronics. These materials are proprietary formulations based on thermoplastic resins, including epoxy, polyester, polyurethane, or acrylic, and may also contain fine aggregate fillers and pigments. Resin floor finishes are applied on-site by specialist contractors, either by trowel (or sledge followed by power trowel), flow applied (self-levelling), or painted on by roller.

**262** Terrazzo floor tile comprising crushed limestone coarse aggregate (large brown particles) and crushed marble fine aggregate particles (white/pink), bound by white Portland cement matrix (brown). The cement matrix is uncarbonated but appears brown owing to optical effects caused by the presence of calcium carbonate dust and pigments; XPT, ×35.

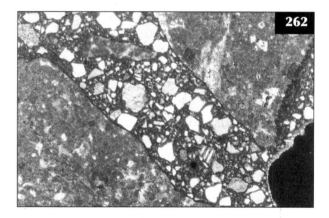

**263** View of the upper surface of terrazzo floor finish showing linear scratches. Aggregate particles appear white and the cement matrix is shown brown; PPR, ×35.

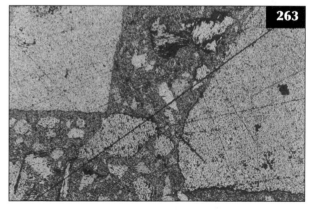

The range of synthetic resin flooring products is shown in *Table 29.*

In thin section, resin binders are isotropic and identification of the resin type is difficult. Although it is theoretically possible to differentiate resins micro-scopically by differences in their refractive index, identification would usually be conducted using an appropriate chemical analysis technique (e.g. infrared spectroscopy). Figure **264** shows a synthetic resin screed flooring (type 6) which has a filler of fine quartz sand and contains no pigment. Figure **265** shows a multilayer synthetic resin floor covering (type 4) which has been laid in three layers. The resin contains a pigment and encapsulates a filler of fine quartz sand, which exhibits settlement.

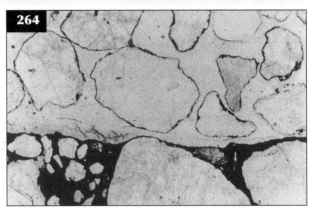

**264** Synthetic resin floor covering (upper, white) consisting of quartz sand filler particles bound by transparent (nonpigmented) resin. This has been laid on a sand:cement screed (lower, dark brown). Air voids are shown yellow; PPT, ×35.

**Table 29** Classification of synthetic resin flooring types (EFNARC, 2001)

| Type* | Class | Details |
|---|---|---|
| 1 | Floor seal | Applied in two coats to give a dry film thickness up to 150 μm; generally solvent- or water-borne |
| 2 | Floor coating | Applied in two or more coats at a dry film thickness up to 100 μm per coat; generally solvent- free or water-bourne |
| 3 | High build floor coating | Applied in two or more coats to give a final thickness of 300–1000 μm; generally solvent-free |
| 4 | Multi-layer flooring | Multiple layers of floor coatings or flow-applied floorings with aggregate dressing, having a thickness >2 mm; often described as 'sandwich' systems |
| 5 | Flow applied flooring | Applied between 2 and 3 mm in thickness. Often referred to as 'self-smoothing' or 'self-levelling' flooring, and having a smooth surface; or may be given surface dressing |
| 6 | Screed flooring | Heavily filled, trowel-finished systems applied at a thickness >4 mm, generally incorporating a surface seal coat to minimize porosity |
| 7 | Heavy duty flowable flooring | Aggregate filled and applied between 4 and 6 mm in thickness and having a smooth surface; or may be given a surface dressing |
| 8 | Heavy duty screed flooring | Trowel-finished, aggregate-filled system applied at a thickness of ≥6 mm or more and effectively impervious throughout its thickness |

\*   These categories are listed in ascending order of durability

Synthetic resin floors do occasionally exhibit defects, with the most common problem being detachment of the resin topping due to poor preparation of the surface on which it has been laid. Other issues involve cracks or ripples reflecting cracks in the substrate, and cracks and collapsed areas ('elephant's footprints') related to failure of the underlying screed. Figure **266** shows a multilayer synthetic resin floor, which has cracked due to collapse of the underlying sand:cement screed. This resin flooring is a type 4 system that incorporates pigment and fine quartz sand filler.

Osmotic blistering is a more unusual defect that involves the formation of water-filled blisters on the surface of the resin floor, where moisture has migrated by osmosis and become trapped. Osmotic blisters are usually associated with self-levelling types of resin floor (Pye & Harrison, 1997). The different types of synthetic resin used for resin floorings have differing ranges of durability to temperature conditions and chemical attack. It should be noted that all of the synthetic resins will break down in the presence of organic solvents and polyester resins are susceptible to attack by alkaline solutions.

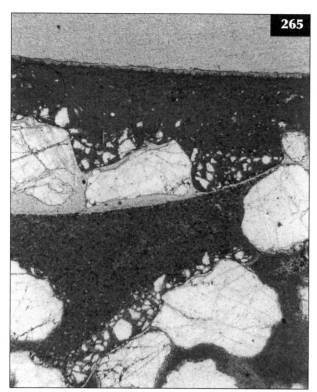

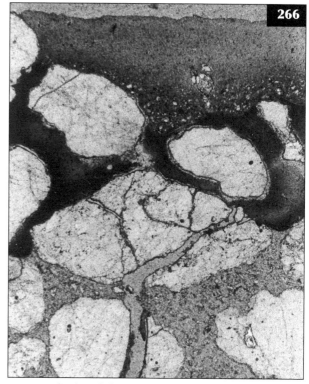

**265** Section through multilayer synthetic resin floor covering, showing three separate layers consisting of quartz sand filler particles (white) bound by pigmented resin (grey/brown). The aggregate within the upper two layers appears to have settled out with the coarsest particles sinking to the bottom; PPT, ×35.

**266** Cracked multilayer synthetic resin floor covering. The covering has cracked under load due to the underlying screed being unacceptably weak. Quartz filler particles appear white, pigmented resin appears grey/brown, and the cracks are shown yellow; PPT, ×35.

# Mortar, plaster, and render

## OVERVIEW

Mortars are used to embed masonry units as well as being used as wall finishing materials internally (plaster) and externally (render). Mud was the first material used for these purposes and, although it is still used in earth construction, it is neither strong nor durable. To satisfy the demand for dependable structures, stronger and more durable binders have evolved, the main ones being based on gypsum, lime, and cement. A brief comparison of the properties of these different binders is provided in *Table 30* and their development is summarized overleaf.

**Table 30** Comparison of the properties of different binders used for mortars, plasters, and renders

| Property | Mud/clay | Gypsum | Lime | Portland cement |
|---|---|---|---|---|
| Heating temperature for production | Air drying | 130–170°C | 850–1200°C | 1250–1500°C |
| Construction speed | Moderate | Fast | Moderate | Fast |
| Strength | Very low | Low | Medium–high | Very high |
| Breathability | High | High | High | Low |
| Shrinkage | High | Low | Medium | High |
| Fire resistance | High | High | Medium | Medium |
| Appropriate applications | Earth structures | Internal plastering | Conservation of historic buildings<br><br>Increasing revival in use for new buildings (does not require movement joints) | Modern masonry structures with movement joints<br><br>Masonry carrying large loads<br><br>Extreme exposure conditions |
| Inappropriate applications | All other structures | External works (except for conservation of existing external gypsum mortars) | Extreme exposure conditions | Conservation of historic buildings |

As it is produced by heating at relatively low temperatures, it is likely that gypsum was the earliest deliberately manufactured cement, although exactly when it was first used is not known. It was frequently used in Egypt during Pharaonic times, extending back at least 4000 years, and it is found in the construction of the Giza pyramids. When used as mortar, gypsum is relatively weak and has limited adhesive power. As testament to this, the Egyptians often used gypsum mortars more as a means to slide large stone blocks into place, rather then to set them together (Arnold, 1991). A drawback of gypsum is that it is water soluble and therefore not suited to external use, particularly in wet climates. Gypsum plaster was not introduced into Britain until the 13th century when it was brought over from France as plaster of Paris (Davey, 1961). Gypsum plasters were then used sporadically as an alternative to, or as an addition to, lime plaster. In the 20th century, gypsum became the material of choice for internal plaster applications in the developed world. Gypsum plasters are perceived to have a number of advantages over lime, including easily controlled setting time according to function, less time delay between application of coats, they do not shrink on drying, and have excellent fire resistance.

A major development in binder technology was the introduction of lime mortars with greater strength and durability than gypsum. The earliest known examples comprise lime wall and floor plasters from Neolithic sites in the Near East dating from 7000–6000BC (Hughes & Válek, 2003). Construction with lime spread through the Middle East, North Africa, Europe, and the Far East, with lime technology apparently developed independently in Central America by the Mayans at around 1000BC (Hansen, 2005). The Greeks and Romans refined the use of pozzolanic additives to lime, intentionally producing hydraulic lime mortar for underwater works (Vitruvius, trans 1960). In the UK, lime:sand mortars, plasters, and renders using nonhydraulic or feebly hydraulic lime as the binder proved to be durable over many centuries and were used routinely until the late 19th century.

Natural cements are made from deposits of calcium carbonate rich in clay minerals, such as septarian nodules. They are quick setting, very hydraulic, and were popular for use in civil engineering and stucco work in 19th century Britain (Kelsall, 1989). Portland cement, patented in Britain in 1824 is produced by 'over burning' a mix of calcium carbonate (limestone or chalk) and an aluminosilicate (clay or shale) forming a very hydraulic cement clinker.

With the development of natural and then hydraulic cements, lime mortars became uneconomic from the builder's viewpoint, as they set and harden slowly, and were perceived to be unsuited for use in wet situations. Cement gauged or pure cement mortars with their rapid hardening and high strength properties then superseded lime mortars and renders. Portland cements are now the dominant binder for masonry mortars and external renders in new buildings of the developed world.

Despite the phenomenal growth in its use, the generic Portland cement has a number of shortcomings, restricting its usage in certain ways. A wide range of other ingredients has been combined with Portland cement to improve its performance when used for special applications. Specialist mortars have been developed for a plethora of special circumstances, including concrete repair, grouting, tiling, mortar jointing, rendering, self-levelling floor finishes, and water-leak stoppers. These tend to be complicated premixed formulations containing multiple ingredients that are manufactured commercially using confidential mix designs. In the 20th century, cement-rich mortars were used in the repair and restoration of historic buildings, where they had most often not been part of the original fabric. The resulting lack of compatibility between dense, impermeable cement-based repairs and traditional construction built in, and protected by, lime-based materials has resulted in an unacceptable level of damage to historic buildings. In the last few decades there has been a renewal of interest in the use of lime, owing to recognition of its superior properties of vapour permeability (breathability), flexibility, and appearance (Induni, 2005). It is now recognized that repair of historic masonry structures should be carried out with a traditional craft-based approach, ideally using materials similar to those used originally to ensure compatibility (British Standards Institution, 1998).

Since the 1990s there has been a modest revival of use of lime mortars for new masonry buildings in the UK. Using hydraulic lime enables rapid construction and does away with the movement joints needed to cope with the shrinkage of strong Portland cement mortars (Beare, 2004). With climate change becoming a global concern, lime binders also have the potential to regain ground as they are more environmentally friendly (Pritchett, 2005).

# GYPSUM-BASED

## INTRODUCTION

Gypsum-based plasters are produced by heating gypsum at relatively low temperatures (130–170°C), driving off three-quarters of its water to form hemihydrate as follows:

$$2(CaSO_4.2H_2O) + heat \rightarrow 2(CaSO_4.\frac{1}{2}H_2O) + 3H_2O$$

Gypsum                    Hemihydrate

Hemihydrate is also known as bassanite or plaster of Paris and it is suitable for a use in a wide range of plaster products. Heating gypsum at temperatures in excess of 400°C causes all of the water to be driven off to form anhydrite as follows:

$$CaSO_4.2H_2O + heat \rightarrow CaSO_4 + 2H_2O$$

Gypsum                 Anhydrite

Anhydrite is restricted to use in wet plastering applications only. Alternatively, crushing natural anhydrite rock will produce plaster directly without heating.

In use, water is added to the hemihydrate and/or anhydrite, which then rehydrates to a mass of set gypsum crystals (the reverse of the above reactions). Calcium sulfate plasters normally set very rapidly and chemical additives are often added to retard the setting time.

For internal plastering, a number of different bagged gypsum products is available to cope with range of internal surfacing applications. Premixed lightweight plaster comprises retarded hemihydrate with a lightweight aggregate. Retarded hemihydrate plaster is used as finishing plaster or, with the addition of sand, as a heavy undercoat plaster. A great deal of hemihydrate production is consumed by production of plasterboard (wallboard) for dry lining internal surfaces, providing a smooth and rigid background upon which a plaster finish can be applied. Plasterboard is hydrated plaster compressed between two layers of strong paper.

Typical applications of petrographic examination to investigation of gypsum plaster in service include:

- Determining the number of layers and their thickness.
- Identifying the type and source of the plaster.
- Assessing the effectiveness of the manufacturing process.
- Identifying the presence/type of impurities and intentional additions.
- Identifying the causes of failures.
- Diagnosing decay mechanisms and assessing the level of deterioration.

## PETROGRAPHIC EXAMINATION AND COMPLEMENTARY TECHNIQUES

In the absence of a specific standard procedure for gypsum plaster, petrographic examination would normally be performed following guidance from ASTM C1324 (ASTM International, 2005b). Once a sample is received in the laboratory, an initial visual and low-power microscopical examination is conducted to determine the number of material types or layers, layer thickness, colour, and the relative hardness, relative bond strength, and coherence of the material. High-power examination of thin section specimens is then used to determine the nature of the gypsum plaster matrix, the quantity of impurities and additions, bond surface characteristics, and outer surface features, and to screen the material for evidence of distress or deterioration.

As gypsum plasters are very fine-grained, the level of confidence in ingredient identification and contaminant detection is greatly increased if chemical analysis is performed in addition to optical microscopy. Suitable techniques include XRD and SEM. The gypsum crystals that dominate the matrix of gypsum plasters can barely be resolved by the highest magnifications of optical microscopy. Therefore, observation using SEM is required for detailed studies of crystal morphology.

## PLASTER PRODUCED FROM NATURAL GYPSUM SOURCES

Pure gypsum is a white to transparent mineral, but impurities can cause it to have a grey, brown, or pink tint. Occurring naturally as an evaporite mineral, gypsum is often associated with 'red-bed' sediments and may be contaminated by haematitic clays. In Britain, gypsum from natural sources yields a plaster that is pink in appearance, due to the persistence of finely disseminated haematite. There are few constraints on the purity of gypsum required for plaster products and standards specify a hemihydrate content as low as 85% in the plaster building product (although they usually contain over 95% hemihydrate). The remainder may comprise impurities including calcium carbonate, clay, and, more rarely, organic matter.

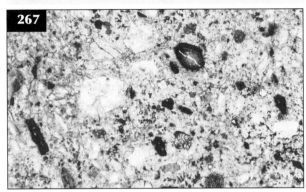

**267** Unhydrated gypsum plaster from a natural source comprising particles of hemihydrate (light brown/white), clay-rich particles (reddish-brown), and one particle of limestone (grey, lower right). The plaster powder was consolidated by epoxy resin (yellow) to allow thin section making; PPT, ×35.

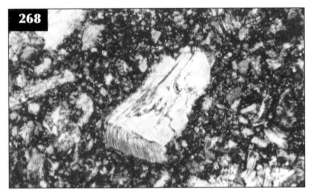

**268** Close view of unhydrated gypsum plaster produced from a natural source, showing a hemihydrate particle (centre, yellow grey) and an anhydrite particle (lower right); XPT, ×150.

Figures **267** and **268** show the typical appearance of British natural source gypsum plaster as supplied by the manufacturer (i.e. unhydrated). It is a fine powder consisting chiefly of hemihydrate (bassanite) particles that appear fibrous and porous, colourless to light brown and exhibit first-order grey to yellow interference colours in cross-polarized light. Occasional particles of colourless anhydrite have a characteristic blocky crystals and strong birefringence (colourful appearance in cross-polarized light). Minor proportions of iron-rich clay impurities appear red and limestone impurities are sporadically seen.

Figures **269** and **270** show the appearance of hydrated gypsum plaster from the same natural British source, chiefly comprising a mass of small, interlocking crystals of gypsum with first-order grey interference colours. Larger gypsum crystals are seen filling voids. Iron-rich clay impurities are still present and clearly identifiable by their red colour. Anhydrite particles still persist as some of them were insoluble (dead burned) and consequently failed to hydrate.

### FLUE GAS DESULFURIZATION GYPSUM PLASTER

An alternative source of gypsum for plaster manufacture is provided by flue gas desulfurization plants installed in coal-burning electricity generating stations. Sulfurous gases scrubbed from power station emissions to reduce pollution, are neutralized with ground limestone to produce flue gas desulfurization gypsum (FGD gypsum) in huge quantities. Apart from a finer grain size, which causes some handling problems, FGD gypsum has proved suitable for hemihydrate production.

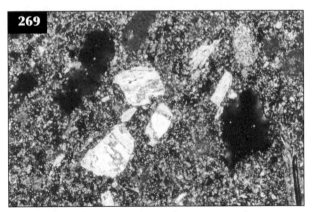

**269** Hydrated gypsum plaster from a natural source, showing anhydrite particles (brightly coloured) with high-order interference colours and iron-rich impurities (red); XPT, ×150.

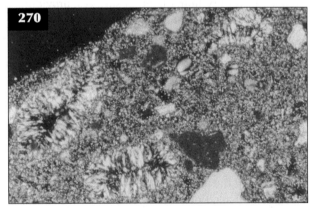

**270** Hydrated gypsum plaster produced from a natural source showing gypsum crystals (grey) lining voids; XPT, ×75.

Figure **271** shows the microscopical appearance of an unhydrated FGD gypsum plaster. The colour of the plaster in hand specimen is white as it consists almost entirely of fine hemihydrate particles and is without the iron-rich impurities of natural source gypsum. Figures **272** and **273** show the appearance of hydrated FGD gypsum plaster from the same source, chiefly comprising of a mass of small, interlocking crystals of gypsum. Traces of limestone and quartz particle impurities are seen in Figure **272** and Figure **273** illustrates growth of larger gypsum crystals in a void.

## AGGREGATES AND ADDITIVES

Low-density aggregates, such as perlite (**274**) and vermiculite (**275**), are important ingredients in modern

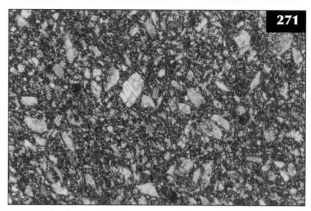

**271** Unhydrated FGD gypsum plaster consisting almost entirely of hemihydrate particles (grey/yellow); XPT, ×35.

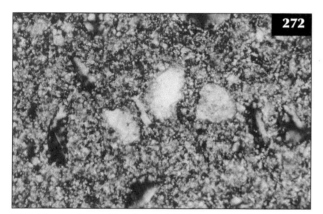

**272** Close view of hydrated FGD gypsum plaster comprising a mass of gypsum crystals (grey) with rare impurities (centre) consisting of limestone (light brown) and quartz (grey) particles; XPT, ×150.

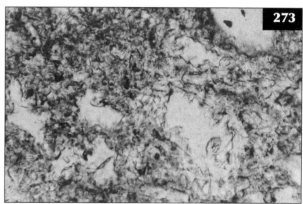

**273** Close view of hydrated FGD gypsum particles showing interlocking crystals of gypsum (white) and air voids. Larger laths of gypsum line the void to the right of centre; PPT, ×300.

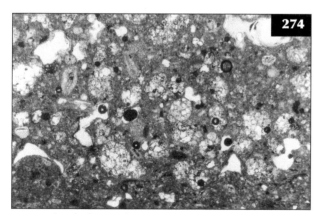

**274** Perlite lightweight aggregate (white) in a premixed lightweight gypsum plaster. The expanded perlite particles comprise clusters of hollow bubbles with walls of volcanic glass; PPT, ×35.

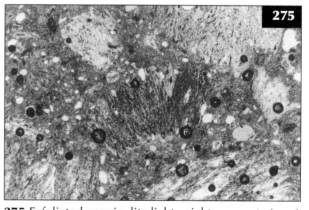

**275** Exfoliated vermiculite lightweight aggregate (grey) in a premixed gypsum plaster (grey), comprising concertina-shaped granules of vermiculite flakes; PPT, ×35.

plasters. Reducing the density of plaster products has a number of advantages including lower transport costs, less effort to mix/apply, it can be used in thicker layers (without sagging), and it has improved thermal insulation and fire resistance properties.

Lime may be used in quantities of up to 25% in certain finishing plasters, to improve working properties of the fresh material and to reduce corrosion of embedded metals (by increasing alkalinity). Undercoat plaster intended for high suction backgrounds might include cellulose fibres to improve water retention.

Historic plasters and those used for restoring historic buildings may be reinforced using animal hair, sisal, straw, or chopped reed (**276**). Modern precast plaster products may incorporate reinforcing fibres such as chopped fibre glass (**277**). Glass fibre and fire retardant additives may also be added to improve fire protection performance.

Chemical additives such as sugar, starch, alum, and potassium/zinc sulfate are added to gypsum plaster as accelerators and retarders. These are not visible microscopically and, as they are usually only present at levels of 0.5–1%, they are also difficult to detect by chemical analysis. It has been found that certain chemical additives can reduce the strength of plaster by detrimentally changing the shape of the set gypsum crystals and their interlocking structure (Prakaypun & Jinawath, 2003). It is also reported that the use of carboxylic acid retarder can improve the durability of repair mixes for historic gypsum-based mortar, by increasing density and lowering capillary porosity (Middendorf, 2001).

## PLASTERBOARD
Patented in America in 1894, the use of plasterboard as an internal dry lining finish is a relatively recent application. It has become popular as an alternative to plaster finishes because it is cleaner and less labour intensive than wet plastering and reduces the drying out time of building construction. Produced by sandwiching accelerated hemihydrate slurry between paper liners and then kiln drying, the strength of plasterboard lies in the heavy paper facing. Gypsum is only required to be low strength inert filler, preferably with low density. Reduction in density is usually achieved by air entrainment, although the incorporation of lightweight materials such as wood fibre and pumice is known. Technological advancements mean that a wide variety of boards is now available to meet an ever-increasing range of applications. Most boards have one surface (often ivory coloured) for direct decoration and one surface (often grey), which has better adhesion properties for

plastering. Other plasterboard varieties include those with foil backing or bonded layers of polystyrene/glass fibre to improve thermal insulation, or a polythene backing for improved vapour resistance. Figure **278** shows the typical aerated texture of plasterboard.

## PLASTER DEFECTS
Petrographic examination is called upon to investigate a variety of issues related to gypsum plasters and associated plastering defects. One of the most common is whether the plaster material used is the one that was specified. This can usually be achieved by microscopical

**276** Particle of chopped reed reinforcement (white) in historic gypsum plaster. Note the cellular plant structure; XPT, ×150.

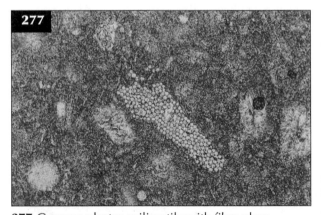

**277** Gypsum plaster ceiling tile with fibre glass reinforcement. The fibre (white) is cut obliquely showing multiple filaments. The gypsum matrix appears light brown and three voids are lined by larger laths of gypsum (white); PPT, ×150.

comparison of the ingredients with a reference sample of the specified plaster. In addition, samples can be screened for impurities that can have a detrimental effect, such as loam and clay which affect the setting time and final strength.

A major drawback of gypsum binders is that they are water soluble and are therefore only suitable for internal work. The absorption of just 1% water can reduce strength by up to half and long-term exposure breaks down the interlocking structure of the gypsum crystals. They were sometimes used externally on historic buildings as mortars and renders, where they usually show signs of severe weathering. Microscopical investigation of historic gypsum mortar weathering typically shows the gypsum binder dissolved away, with the remainder appearing reprecipitated as larger crystals with a denser microstructure (Middendorf, 2001). Observation using SEM is required to study the morphology of the very fine crystals found in gypsum plasters.

Workmanship issues that can be addressed include checking that the plaster has been applied using the specified number of layers, at the recommended thickness. Finishing plaster can debond if the undercoat has not had time to set, if it has an inadequate mechanical key to the undercoat, or if a strong finishing coat is applied to a weak undercoat. Figure **278** shows a finishing plaster that has debonded from its plasterboard background. Gypsum plasters are nonshrinking but may suffer cracking for other reasons when subjected to

movement in excess of their strain capacity. Common causes of cracking include background shrinkage (if the background was very wet), base coat shrinkage (with cement-based undercoats), cracks reflecting plasterboard joints, and building movement. An unsatisfactory soft powdery surface may result if the plaster dries out before it has fully hydrated. This is more likely to happen if thin coats are applied to a dry background, such as plasterboard.

## LIME-BASED

### INTRODUCTION
Lime is formed by burning a source of calcium carbonate ($CaCO_3$), usually limestone or magnesian limestone, between 850°C and 1200°C, driving off carbon dioxide to form calcium oxide (CaO, quicklime). The calcium oxide is then slaked with water (evolving heat) to form calcium hydroxide (lime, $Ca(OH)_2$). The slaking process can produce dry lime hydrate powders or, if excess water is used, lime putty. Lime mortars, plasters, and renders are manufactured by mixing aggregate with a lime product, either quicklime, lime hydrate powder, or lime putty, adding water if required. Lime made from pure limestone (nonhydraulic lime) sets by drying out and then hardens wholly by absorption and reaction with carbon dioxide, slowly to become calcium carbonate again.

*Burning*    $CaCO_{3(s)} + heat \rightarrow CaO_{(s)} + CO_{2(g)}$
*Slaking*    $CaO_{(s)} + H_2O_{(l)} \rightarrow Ca(OH)_{2(s)}$
*Hardening*    $Ca(OH)_{2(s)} + CO_{2(g)} \rightarrow CaCO_{3(s)} + H_2O_{(l)}$

Lime made from limestone containing siliceous impurities has a degree of cementitious properties, allowing it to set underwater (hydraulic lime). This occurs as calcium and silica from within the impure limestone react to form materials similar to those found in Portland cement, such as dicalcium silicate, aluminate, and ferrite phases, when the rock is calcined. The mechanism of hardening of hydraulic limes is a combination of carbonation and hydration of dicalcium silicate.

The principal applications of petrographic examination to investigation of modern and historic lime mortars are:
- Identifying the type and source of aggregate.
- Identifying the type, origin, and manufacturing process of the binder.
- Direct estimation of mix proportions.
- Enabling correction of mix proportions determined by simple chemical analysis.
- Identifying and quantifying the presence of inclusions, additives, and impurities.
- Assessing workmanship.

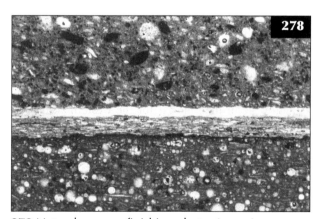

**278** Natural gypsum finishing plaster (upper) debonded from paper-faced plasterboard (lower). The plasterboard comprises gypsum (brown) with abundant entrained air voids (white, spherical). The heavy paper facing is clearly visible running horizontally across the centre of view; PPT, ×35.

- Identifying defects.
- Diagnosing decay mechanisms and assessing the level of deterioration.
- Identifying and measuring existing surface treatments.

## PETROGRAPHIC EXAMINATION AND COMPLEMENTARY TECHNIQUES

Petrographic examination of lime mortar, plaster, and render is performed following a procedure given in ASTM C1324 (ASTM International, 2005b). At the time of writing a European standard was being developed but was not yet available. Once a mortar sample is received in the laboratory, an initial examination is conducted with the unaided eye and low-power binocular (stereoscopic) microscope. Combined with some simple physical and chemical tests this gives a rapid and economical provisional indication of the number of material types present, the number of coats or layers, the particle size distribution (grading), particle shape and mineralogy of the aggregate, the presence of inclusions, and the type and relative hydraulicity of the binder. The colour of both the aggregate and the binder can be classified by comparison with a Munsell™ colour chart (Munsell, 1994).

Although initial visual and low-power microscopical examination gives an indication of mortar ingredients, definitive identification of the aggregate and binder composition always requires more detailed high-power microscopical examination in thin section. Thin section examination can provide a great deal of information regarding the ingredients used, their relative proportions, overall texture, porosity, and condition.

Information gained from optical microscopy can be supplemented/complemented by other microscopical and analytical techniques. Where closer examination is required, or the constituents are too fine to be resolved by optical microscopy, SEM is invaluable, and integral microanalysis systems provide elemental compositions to aid in material identification. A particularly useful technique is environmental SEM which allows examination and analysis of freshly fractured surfaces without any damaging sample preparation (Radonjic *et al.*, 2001a). Environmental SEM has even been used successfully to study changes in lime specimens while they are carbonating (Radonjic *et al.*, 2001b). Mineralogical analysis by XRD can be helpful for the identification of crystalline minerals and decay reaction products when their optical properties do not allow definitive identification. Infrared spectroscopy is the preferred technique to detect and identify the presence of organic additives such as tallow or linseed oil.

## AGGREGATE

The aggregate for lime mortars, plasters, and renders may be derived from different types of geological deposits, including natural sand resources (pit, river, beach, marine dredged) and rock formations (quarried). Natural sands often comprise rounded 'as-dug' particles, but may include a proportion of more angular particles if oversize materials have been crushed down to the required size. Quarried rocks always have to be crushed down to achieve the required particle size/grading and consequently comprise particles that are quite angular in shape. Occasionally, aggregates from more than one source are blended together to achieve desirable overall aggregate characteristics. The surface texture of aggregate particles and the range of different particle sizes represented (grading) have a large influence on the properties and quality of the resulting mortar. Desirable aggregates for modern brickwork mortars are usually uniformly graded with spherical, rounded particles ('soft sand') to aid workability (**279**), while a repair specification for a historic render would require well graded aggregate with irregular, subangular particles ('sharp sand') to reduce drying shrinkage (**280**). Particle shape and grading characteristics can be estimated through the microscope, aided by comparison with standard charts (**130** and **133**). Image analysis techniques are also being developed to determine grading automatically (Mertens & Elson, 2005). The mineralogy of the aggregate reflects its geological origin and gives clues to its source. For example, in southeast England the terrace sands of the River Thames were widely used (**279** and **280**); less widely used were crushed flints from the Chalk (**281**). River gravels derived from sandstone country rock were used by the Romans as aggregate in lime mortar for the western section of Hadrian's Wall in northern England (**282**). Calcareous aggregates were used in areas where it is the most convenient source, for example in Oxfordshire, UK, where the underlying geology comprises Jurassic limestone (**283**). Calcareous shells may often be incorporated in marine dredged or beach sands (**284**).

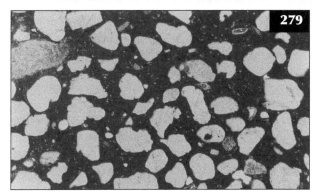

**279** Lime mortar with uniformly graded sand aggregate, showing quartz fine aggregate particles (white) and lime binder (dark brown); PPT, ×35.

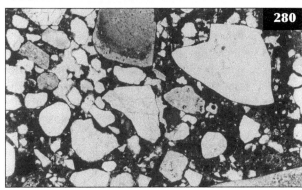

**280** Lime render with well graded sand aggregate, showing quartz (white) and flint (light brown) fine aggregate particles and lime binder (dark brown). Air voids are shown yellow; PPT, ×35. (From Ingham, 2004.)

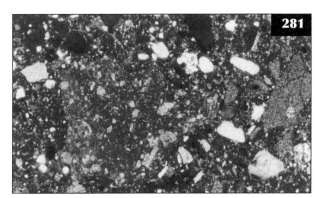

**281** Roman lime plaster from archaeological excavations at Verulamium in southeast England. Showing crushed flint (grey) and natural quartz (white) fine aggregate particles and an unburnt chalk particle (brown). The carbonated lime binder appears dark brown; XPT, ×35.

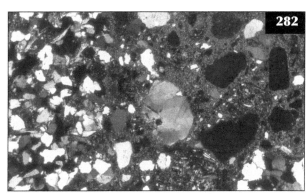

**282** Roman mortar from Hadrian's Wall, showing sandstone (left), quartz (grey/white), and basalt fine aggregate particles, and carbonated nonhydraulic lime binder (brown); XPT, ×35.

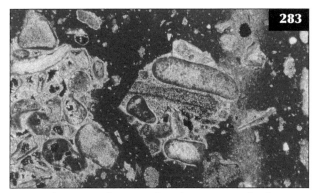

**283** 18th century lime render, showing crushed fossiliferous limestone (light brown) fine aggregate and carbonated lime binder (dark brown); XPT, ×35.

**284** 19th century lime render from a lighthouse, showing fine aggregate composed chiefly of shell fragments. The particle near the centre with radial appearance is an echinoid spine; XPT, ×35.

## LIME BINDERS

Carbonated (hardened) lime binders have a distinctive texture comprising finely crystalline calcite with occasional pockets of coarsely crystalline calcite deriving from the slow carbonation of trapped slaked lime. The relative hydraulicity of the lime can be determined by observing the quantity of small relict grains chiefly comprising dicalcium silicate ('belite') (285). Nonhydraulic limes have very few (if any) of these belite grains, with the number of grains observed increasing through the feebly hydraulic, moderately hydraulic, and eminently hydraulic classes (British Standards Institution, 2000). The binder of historic lime mortars can comprise anything up to about 35% of lime inclusions (Leslie & Hughes, 2002). Usually roughly spherical lumps of 50 μm to 10 mm in size, these inclusions can consist of unmixed lime lumps (286, 298, and 313), typically consisting of underburnt, partially burnt or overburnt quicklime (287 and 288). The presence of large (>3 mm sized) unmixed lime lumps is sometimes interpreted to have been caused by 'hot mixing' or 'dry slaking', a practice where damp aggregate is mixed with roughly crushed quicklime (Hughes *et al.*, 2001). Underburnt or partially burnt inclusions often exhibit the original limestone texture, enabling identification of the geological source of the limestone burnt in the kiln.

If the lime has been gauged with Roman cement or Portland cement this will usually be detected by the observation of unhydrated relict cement grains composed of dicalcium silicate crystals and aluminous and ferrous

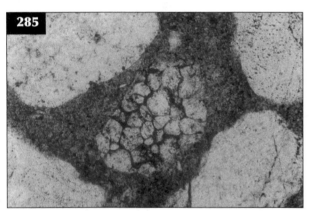

**285** Relict grain of dicalcium silicate (belite, centre) within a 1930s hydraulic lime render. The lime binder appears brown and quartz fine aggregate particles are shown white; PPT, ×300.

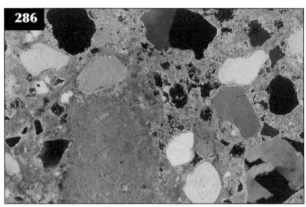

**286** Lump of unmixed lime (light brown) within a lime/Portland cement render. Quartz natural sand fine aggregate particles appear grey/white and two small unhydrated Portland cement grains are right of centre (dark brown); XPT, ×35.

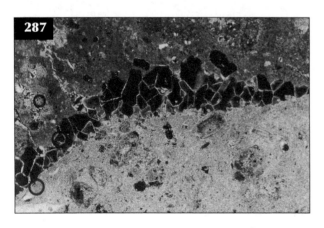

**287** Partially burnt limestone inclusion within Roman lime flooring concrete from archaeological excavations at Piddington Roman Villa, England. There is a burnt rim (dark brown) around the unburnt core (lower, light brown); XPT, ×35.

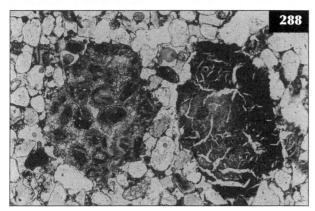

**288** Inclusions of unburnt limestone (left) and an overburnt lime lump (right) within medieval mortar from a Welsh castle; PPT, ×35.

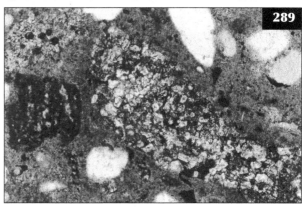

**289** Large relict Portland cement grain (centre) within early 20th century Portland cement/lime render, comprising calcium silicates (white) and interstitial phases (dark brown); PPT, ×150.

interstitial phases, and in the case of Portland cement, also including tricalcium silicate (**289**). These relict cement grains were relatively large (up to 500 µm across) in early hydraulic cement mortars and have decreased in size with time as clinker grinding technology has improved, with modern Portland cement mortars having cement grains of <30 µm in size. Lime plasters were sometimes gauged with gypsum, which is detected optically by the presence of finely disseminated gypsum within the matrix, large gypsum crystals lining air voids, relict particles of hemihydrate (bassanite), and for English gypsum sources, clay-rich lump impurities (**290** and **291**).

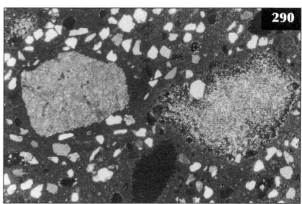

**290** Historic lime/gypsum plaster, showing a large lump of hemihydrate (grey, right) from gypsum and a lump of unburnt limestone from the lime (light brown, left). Fine aggregate consists mainly of quartz (white/grey) and the carbonated binder appears dark brown; XPT, ×35.

**291** An unusual early 20th century plaster comprising lime gauged with both gypsum and Portland cement. The binder has the appearance of carbonated lime (light brown) and encloses clay-rich impurities (red/orange). Gypsum crystals lining air voids indicate the addition of gypsum and relict Portland cement grains are common (small and dark brown); XPT, ×75.

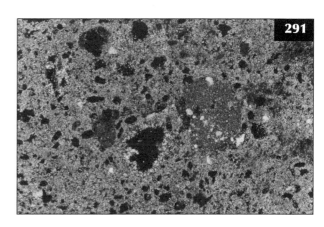

## POZZOLANA

The advantages of pozzolanic additives to lime mortars have been known since antiquity when the Greeks and Romans refined their use, intentionally producing lime mortars for underwater works. The Romans in particular sought out pozzolanic pumice for use in lime concrete for large-scale harbour structures around the Mediterranean. It was thought so beneficial that in many cases they transported them over great distances (Oleson *et al.*, 2004). Examination of these lime concretes shows clear evidence of pozzolanic reaction rims around the pumice particles (**292**). Crushed ceramic brick, tiles, and potsherds were used in Roman lime mortars and concretes for small-scale and terrestrial purposes (Siddall, 2000) (**293**). Brick dust has been added to English lime renders since the 18th century making them dark brown in colour (**294**). The degree of pozzolanic activity of ceramic additives depends on factors such as the firing temperature of the clay and the composition, state, and particle size of the ceramic product, all of which can be investigated microscopically. The presence of other, more modern pozzolanas, such as PFA and trass, is easily detected optically, owing to their distinctive morphologies, while very finely divided mineral additions, such as metakaolin, are too small to be seen using optical microscopy alone. The detection of very finely divided mineral additives relies on observation of indirect evidence such as high optical density of the binder and the presence of agglomerations of undispersed material.

## INCLUSIONS

Impurities such as charcoal (**295**) or wood are often observed in historic mortars. Small quantities of charcoal usually originate accidentally as remnants of fuel from the lime kiln. If large quantities of charcoal or furnace clinker are present, then they were probably added deliberately to colour the mortar black. Small inclusions of wood may represent fuel remnants or possibly contamination by fragments of the mortar board or plasterer's hawk.

Lumps of finely crystalline sparry calcite may be present in lime mortars (**296**). These may originate as a hard carbonate crust that can form on top and around the edges of lime putty or 'coarse stuff' as it matures in pits or other containers. This carbonated crust was often accidentally or deliberately crunched into the mix prior to application and is believed to aid the overall carbonation process.

Animal hair has historically been used as reinforcement for lime plasters (**297**). Typically, around 4–8 kg of hair was added to a cubic metre of plaster and the hair would ideally be strong, soft, not too springy, and of various lengths between 25 mm and 100 mm. Goat and cattle hair are accepted as the best varieties, but less suitable horse and even human hair have been used historically. As a rule of thumb, goat hair is light coloured (commonly white), while cattle hair is usually dark (commonly brown or reddish-brown) and horse hair can be a variety of colours. Goat and cattle hair are covered by tiny barbs which hold the hairs in place within the mortar. Horse hair has a smooth surface and consequently can be pulled away from the hardened mortar by hand. Human hair was rarely used because of its fineness and poor strength. It should be noted that it is theoretically possible to contract anthrax from infected hair within historic plaster, although the risk is extremely low (English Heritage, 2001). Simple contact reduction precautions and good hygiene practices are advisable when handling historic plaster specimens.

## MIX PROPORTIONS

Modern, well mixed lime mortars typically have mix proportions of around 1:3 (lime:sand). Historic lime mortars often contain more lime with mix proportions of 1:1 or 1:2. However, historic mortars typically contain more lime inclusions in the form of lumps of unmixed lime, underburnt material, or hardburnt particles. Although derived from the lime binder these inclusions do not function as binder in the mortar and in terms of performance can be regarded as a form of aggregate (Leslie & Hughes, 2002). However, for the purposes of matching, it is the original binder content including the lime inclusions which is used, so the repair lime should include similar types/amounts of inclusions (Ingham, 2003). In rare cases lime mortars were used historically with no aggregate at all. For example, Figure **298** shows

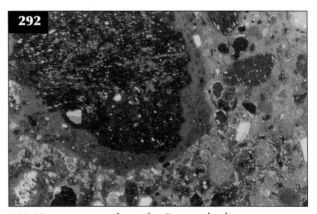

**292** Lime concrete from the Roman harbour at Corinth, Greece, showing a pumice aggregate particle with a pozzolanic reaction rim (appears darker than the carbonated lime binder); XPT, ×35.

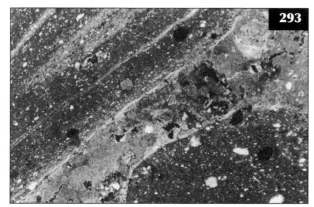

**293** Ceramic tile (red) used as aggregate within Roman lime flooring concrete from archaeological excavations at Piddington Roman Villa, England; XPT, ×35.

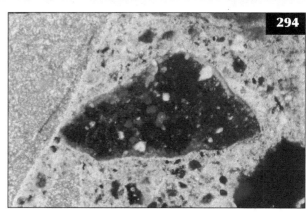

**294** A brick dust inclusion (red) within a historic lime mortar. Flint fine aggregate particles appear grey and the carbonated lime binder appears light brown; XPT, ×150.

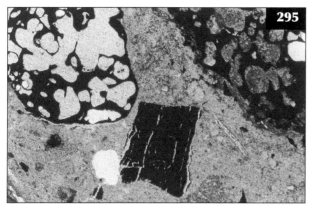

**295** Inclusions of charcoal (black) within 17th century lime mortar; PPT, ×75.

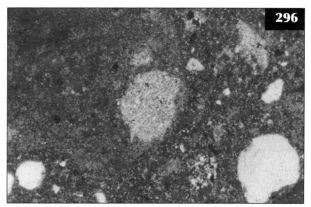

**296** A particle of sparry calcite (light brown, centre), probably originating as a pit crust inclusion from Roman flooring concrete from archaeological excavations at Verulamium in southeast England (Ingham, 2004); XPT, ×150.

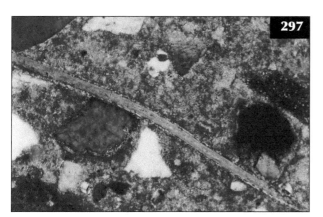

**297** Cattle hair (orange) within lime plaster from a 19th century building; XPT, ×150.

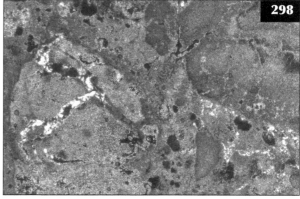

**298** Lime mortar from the Great Wall of China consisting wholly of carbonated nonhydraulic lime with numerous unmixed lumps; XPT, ×35.

Ming Dynasty lime mortar from the Great Wall of China, which consists almost entirely of nonhydraulic lime.

Mix proportions of lime mortars, plasters, and renders can be determined directly from thin sections by three methods. The first method involves visual comparison with reference thin sections of mortar with known mix proportions. The second involves determination of the area occupied by each mortar constituent (aggregate, lime binder, lime lumps, and air voids) by point-counting and then calculating the mix proportions using assumed densities (RILEM, 2001). The third method is a variation of the second in that the areas occupied by the various mortar constituents is determined by digital image analysis. In addition to the binder to aggregate ratio, digital image analysis potentially can be used to determine the particle size distribution and shape parameters of aggregate (Carò et al., 2006) and porosity (Stefanidou, 2000). For digital image analysis techniques it is usually necessary to increase the contrast between the constituents by staining the binder (e.g. by using alizarin red) or by impregnation of pores with coloured epoxy resins (Lindqvist & Johansson, 2007). For all of the above methods care should be taken to ensure that the plane of thin section is as representative of the whole material as possible.

Mix proportions are commonly determined by simple chemical analysis involving acid dissolution, for example in accordance with the BS 4551 method (British Standards Institution, 2005a) or the ASTM 1342 method (ASTM International, 2005b). These methods have a number of disadvantages, the main one being that carbonate aggregate particles (limestone, marble, shell) will be dissolved away with the binder, giving incorrectly high binder content values. Values of soluble silica used for the estimation of the degree of hydraulicity of the lime binder are subject to error if pozzolanas are present, owing to the presence of soluble silica reaction rims around pozzolana particles. Estimation of the type and quantity of calcareous aggregate, lime lumps, and pozzolana made through the microscope can be used to correct erroneous results of wet chemical analysis.

## WORKMANSHIP

Workmanship issues that can be addressed by petrographic examination include the compliance with specified ingredients, mix proportions, number/thickness of coats or layers, adequacy of mixing, effectiveness of compaction, and control of setting conditions.

If plaster or render layers are applied too thickly, or are allowed to dry out too rapidly they are prone to drying shrinkage cracking. Thick layers of lime mortar will also take longer to carbonate and gain strength/hardness. A common cause of lime mortar weakness is deficiency of binder owing to poor batching practices, or inadequate mixing that can result in lean patches of weak material (**299**).

Poorly compacted mortar, plaster, or render with a high content of entrapped air voids will be weaker than anticipated (**300**). The volume of most forms of macropores in addition to their size and shape, are

**299** Weak lime render with a deficiency of binder. Fine aggregate particles appear white, lime binder appears brown, and air voids are shown yellow; PPT, ×35.

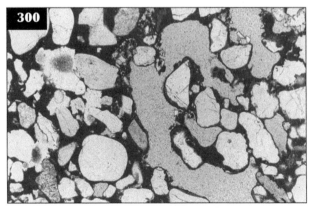

**300** Lime render exhibiting poor compaction of the float coat (right) on to the scratch coat (left). Fine aggregate particles appear white, lime binder appears brown, and air voids are shown yellow; PPT, ×35.

readily observed in thin section. The quantity of large (>1 mm diameter), irregular-shaped entrapped air voids is estimated to assess the effectiveness of compaction at placement. Small (1 mm to 5 µm diameter), spherical air voids are entrained in the mix for a number of reasons and an abundance of such air voids may indicate the use of air entraining (surface-active) or plasticizing (water-reducing) admixtures. The abundance of micropores (defined arbitrarily as being <5 µm across) and therefore microporosity, can be assessed by fluorescence microscopy of thin sections impregnated with resin containing fluorescent dye.

The hardening of nonhydraulic lime mortar by carbonation is a delicate process dependent on temperature, moisture, layer thickness, and pore structure, the properties of the background, and the presence of carbon dioxide. Consequently, careful control of the setting conditions is crucial to successful lime work. The setting of nonhydraulic lime depends on the absorption of carbon dioxide from the atmosphere but also requires the presence of some moisture. The calcium hydroxide ($Ca(OH)_2$) combines with excess carbon dioxide ($CO_2$) to form calcium bicarbonate ($Ca(HCO_3)_2$), which decomposes on evaporation to form crystals of calcium carbonate ($CaCO_3$). The moisture ($H_2O$) helps to dissolve the next particle of calcium hydroxide, forming a saturated solution and putting it into a condition to take up a molecule of carbon dioxide. This, in turn, repeats the action already described and crystals of calcium carbonate are formed (Swallow & Carrington, 1995):

$$Ca(OH)_{2(s)} + CO_{2(g)} \rightarrow Ca(HCO_3)_{2(s)}$$
$$Ca(CO_3)_{2(s)} \rightarrow CaCO_{3(s)} + H_2O_{(l)} + CO_{2(g)}$$

Rapid drying out can occur in hot weather, on unprotected work or if the background is of inappropriate type or too dry. This can result in weak and friable lime mortar/render that has failed to carbonate. Nonhydraulic limes can also fail to carbonate during cold weather conditions, leaving them susceptible to weathering by frost and driving rain. Determination of the degree of carbonation (including partial carbonation) can be achieved definitively by observation in thin section. Rapid drying can also cause plastic shrinkage cracks that are observed microscopically as distinctive tension gashes (**301**).

## PHYSICAL AND CHEMICAL DETERIORATION

Microscopical examination is eminently suited to screening for evidence of decay caused by leaching, salt attack, freeze–thaw damage, and, for mortars including Portland cement, sulfate attack. Signs of matrix replacement/recrystallization, the presence of secondary deposits, and the presence of filled or unfilled cracks/microcracks may provide evidence of deterioration caused by deleterious reactions.

Excessive leaching of masonry by moisture movement can dissolve away the binder within mortar with consequent loss of strength (**302**). Calcium carbonate leached out of lime binder may be redeposited in voids and cracks within the mortar or onto external surfaces.

**301** Lime render exhibiting tension gashes (shown yellow) caused by rapid drying. Fine aggregate particles appear white and lime binder is shown brown; PPT, ×35.

**302** Severely leached medieval lime mortar from a castle. Fine aggregate particles appear white, lime binder appears brown, and air voids are shown yellow; PPT, ×35.

Occasionally these secondary leaching deposits may fill or 'heal' open cracks within the mortar and, in effect, repair the damage. This is referred to as 'autogenous healing' (**303**).

High levels of sulfate can result from the leaching of sulfate-rich inclusions (e.g. charcoal), adjacent clay brick masonry or soil, or polluted masonry surfaces, to cause the expansive formation of gypsum (calcium sulfate hydrate). The expansive growth of gypsum and other salts can cause cracking and other deterioration (**304**).

Lime mortars are poor conductors of heat but nevertheless can suffer considerable damage when exposed to fire. The effects of fire damage can include discolouration, surface cracking, surface spalling, and calcination of the lime binder. Optical microscopy can contribute greatly to the assessment of fire-damaged masonry by enabling the depth of damage to be ascertained, prior to repair. Colour changes and other features associated with heating enable thermal contours (isograds) to be plotted at various depths from the outer surface. For example, iron compounds present within some aggregate particles will oxidize in the range 250–300°C to give a pink to red colour (**305**). The nature and depth of cracking associated with fire damage can be investigated by fluorescence microscopy.

### SURFACE FINISHES

Masonry surfaces have long been painted to improve their appearance and durability. The earliest surface coatings included water-based solutions of lime (limewash), to which natural oils or glues could be added to improve adhesion and reduce chalking (**306**). Natural pigments were sometimes added to the limewash to achieve a desired colour, or applied to damp lime plaster, as in fresco painting (**307**). Other traditional paints comprised natural and manufactured pigments with an oil-based binder. Synthetic paints with plastic-based binders were introduced in the 1930s. Traditional coatings such as limewash are relatively permeable and protect the wall while allowing it to breath and dry out. Modern synthetic paints are often impermeable and while they do provide external weatherproofing, they do not allow dampness trapped within the masonry to dry out. In such situations the weatherproofing paint can itself be the cause of moisture-related decay, owing to leaching and deposition of salts beneath the paint coating (**308**). When investigating surface coatings, specimens are routinely examined microscopically to determine the number and type of layers present and to measure their thickness.

# PORTLAND CEMENT-BASED

## INTRODUCTION

Portland cement is a highly processed form of artificial hydraulic lime, which hardens rapidly by hydraulic reaction alone to form a binder chiefly comprising calcium silicate hydrate gel. The formation of calcium silicate hydrates during the hardening of cements results in stronger, denser, and less permeable mortars than when lime is used as a binder. Mortars and renders based on Portland cement have been the norm for new building since the 1930s. They are produced by mixing aggregate with unhydrated Portland cement and adding only sufficient water to hydrate the cement and provide adequate workability. Mortars are either mixed on-site or they are supplied to site as ready-to-use products. Site mixing of mortar requires careful planning and execution to ensure that a satisfactory material is produced consistently. Ready-to-use retarded cement mortars reduce the problems associated with site mixing operations, such as inaccurate batching and colour variation. Chemical admixtures such as plasticizers, retarders, and air entrainers are commonly added to Portland cement mortars to aid workability, increase setting time, and improve durability.

The main applications of petrographic examination to investigation of Portland cement-based mortars and renders are:
- Determining the number of layers and their thickness.
- Identifying the type and source of aggregate.
- Identifying the type of binder present.
- Identifying the presence/type of mineral additions and pigments.
- Enabling correction of mix proportions determined by chemical analysis through identifying and quantifying the materials present.
- Assessing workmanship.
- Identifying defects.
- Diagnosing decay mechanisms and assessing the level of deterioration.

## PETROGRAPHIC EXAMINATION TECHNIQUES

Petrographic examination of Portland cement-based mortar and render is performed following a procedure given in ASTM C1324 (ASTM International, 2005b), which is largely based on ASTM C856 (ASTM International, 2004). At the time of writing a European standard was being developed but was not yet available. Once a sample is received in the laboratory, an initial visual and low-power microscopical examination is conducted to determine the number of material types or layers, layer

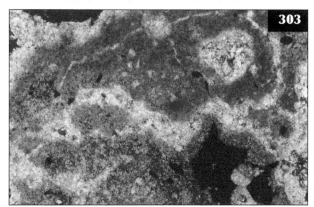

**303** Secondary deposits of coarse calcite crystals (pink) lining cracks and air voids within 17th century lime mortar; XPT, ×150.

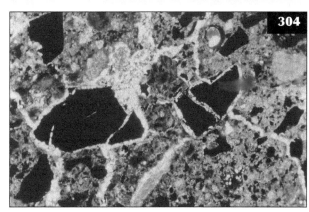

**304** Cracks filled with gypsum (grey) within lime mortar containing sulfate-rich furnace clinker aggregate (black). Lime mortar from a mine building; XPT, ×75.

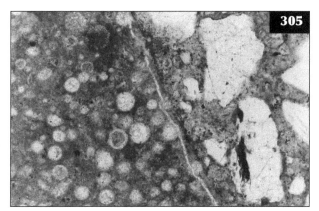

**305** Red discolouration of chalk aggregate particle (left) within fire-damaged lime mortar; PPT, ×150.

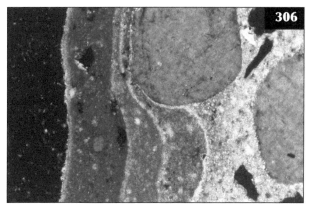

**306** Three layers of limewash coating lime render, as seen 5 years after application. Note the variation in limewash thickness; XPT, ×150.

**307** Decorated Roman wall plaster from archaeological excavations at Verulamium in southeast England (Ingham, 2004), showing three limewash coats with the outer including a yellow pigment (probably ochre); PPT, ×300.

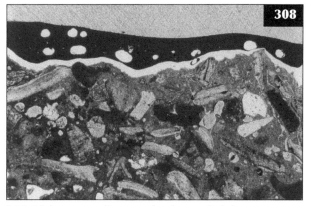

**308** 19th century lime render from a lighthouse. Impermeable modern paint coating (black) over historic lime render. Salts have become trapped behind the paint (white); PPT, ×35.

thickness, colour of aggregate/binder, and the relative hardness and coherence of the material. High-power examination of thin section specimens is then used to determine the grading and mineralogy of the aggregate, cement type, the presence of mineral additions/pigments, assess the quality of workmanship, and screen the material for evidence of distress or deterioration.

## AGGREGATE

The aggregates used for Portland cement-based masonry mortars are usually uniformly graded with spherical, rounded particles ('soft sand') to aid workability (**279**). This contrasts with that preferred for external render which should be well graded aggregate with irregular, subangular particles ('sharp sand') to reduce drying shrinkage (**280**). The aggregate grading, particle shape, and surface texture are important factors affecting the durability of the hardened mortar and these are assessed in thin section by comparison with standard charts (e.g. **130** and **133**). Figure **309** shows the typical appearance of Portland cement:sand mortar with a moderately well graded, natural sand fine aggregate. The mineralogy of the aggregate dictates the strength and soundness of the aggregate particles. Aggregate may contain unacceptable amounts of potentially deleterious materials, including mica (**310**) or excessive fines, which may cause the mix to be weaker than intended, by increasing the water demand. Other undesirable aggregate constituents include pyrite,

salts, or organic matter that may cause unsoundness. The colour of the aggregate and, in particular, aggregate particle coatings and fines may have a significant influence of the overall colour of the hardened mortar. Figure **311** shows a mortar that has been coloured pink by iron oxides within the fine aggregate.

## CEMENT BINDER

Portland cement binders for mortars and renders have a similar appearance in thin section to the cement matrix of hardened concrete. Figure **309** shows a typical uncarbonated Portland cement:sand mortar with an isotropic matrix of calcium silicate hydrates including dispersed portlandite (calcium hydroxide) crystallites (first-order red and second-order blue interference colours in cross-polarized light). Unhydrated/hemi-hydrated cement grains are usually easy to observe in plane-polarized light at magnifications of greater than ×100 (**312**). Observation of the phases present within unhydrated cement grains enables identification of the type of cement used. The size and spacing of relict cement grains are a function of fineness of the ground cement clinker, the W/C of the mix, and the degree of hydration.

Portland cement:sand mixes tend to be rather harsh and it is customary to use either lime or air-entraining agents to add cohesion and aid workability. Figure **313** shows the texture of an old Portland cement:lime:sand render. The

**309** General view of a Portland cement:sand mortar from London, showing quartz (grey/white), flint (grey, elongated), and glauconite (green/yellow) fine aggregate particles (Thames Valley sand), bound by a matrix of uncarbonated Portland-type cement (black). The cement matrix exhibits a common abundance of portlandite crystallites (brightly coloured 'specks'); XPT, ×35.

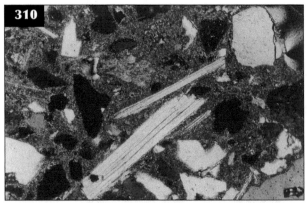

**310** Portland cement:sand mortar from southwest England, comprising crushed granite fine aggregate consisting of angular quartz (white/grey/black), feldspar (grey), and muscovite mica (bright green/blue flakes) particles, bound by a matrix of carbonated Portland-type cement (brown). Air voids are shown dark green; XPT, ×75.

carbonated cement/lime binder has a similar appearance to that of carbonated cement paste and it would be easy to overlook the presence of lime if it were not for the sporadic occurrences of small unmixed lime lumps. The presence of relict Portland cement grains confirms the use of Portland cement as a binder ingredient. In well mixed, fully carbonated Portland cement:lime:sand mixes it is difficult to detect the lime, and chemical analysis must be relied upon to confirm its use.

Mortars for masonry jointing may use masonry cement to produce a more workable and cohesive mortar than using ordinary (CEM I or ASTM Type I) Portland cement. Masonry cement comprises Portland cement finely interground with limestone or another inert filler, often used in combination with plasticizing or waterproofing chemical admixtures. As a significant proportion of the cement has been replaced with inert filler, mortar mixes based on masonry cement should have a lower proportion of sand compared with mixes made with other cements, in order to achieve similar proportions of cement to sand. Figure 314 shows the texture of masonry cement mortar, notable for the frequent abundance of limestone dust filler visible within the cement matrix.

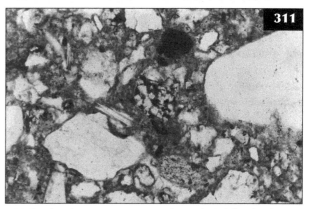

**311** Portland cement mortar coloured pink by the fine aggregate. The fine aggregate comprises mainly quartz particles (white) with a thin veneer of red iron oxide and includes a proportion of iron-rich fines. An unhydrated cement grain is also seen (centre); PPT, ×150.

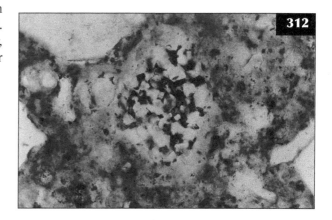

**312** Close view of an unhydrated cement grain (centre) within a Portland cement-based mortar. It comprises calcium silicates (white) and interstitial phases (dark brown); PPT, ×300.

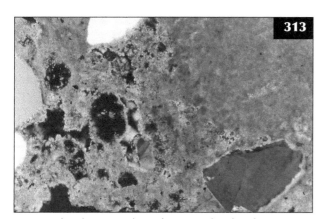

**313** Portland cement-based mortar that has been gauged with lime to improve workability, showing an unmixed lump of lime (upper right) and carbonated cement/lime matrix (light brown) that includes sporadic unhydrated Portland cement grains (dark brown). Quartz fine aggregate particles appear grey/white; XPT, ×150.

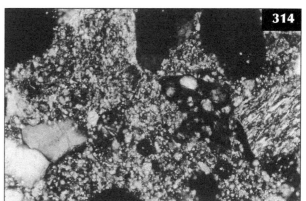

**314** Close view of a masonry cement blockwork mortar, showing a frequent abundance of limestone dust filler (pink/yellow) within the uncarbonated cement matrix. There are natural quartzitic sand fine aggregate particles (grey/white). An unhydrated cement grain is seen centre right; XPT, ×300.

To prevent sulfate attack, mortars made with sulfate-resisting cement (**167** and **170**) may be used with bricks manufactured with high soluble sulfate contents. They may also be used for chimneys where sulfate from flue gases can be deposited, or masonry in contact with sulfate-rich soils or groundwater.

Ordinary Portland cement hardens to a grey colour and, for aesthetic reasons, white Portland cement is sometimes used as an alternative (**315**). Figure 316 shows a case where white Portland cement has been mixed with light brown, limestone fine aggregate to achieve an overall buff-coloured decorative render. Figure **317** shows an example of joint mortar used between blocks of

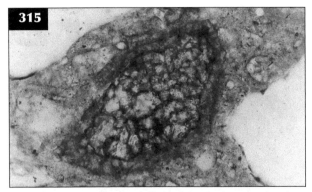

**315** Close view of a partially hydrated white Portland cement grain in a decorative render. The grain consists of calcium silicates with no interstitial ferrite; PPT, ×300.

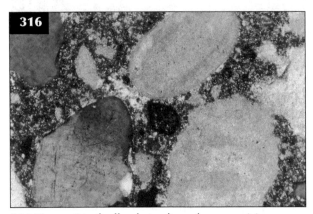

**316** Decorative (buff-coloured) render comprising mixed crushed brown limestone (light brown) and natural quartz-rich sand fine aggregate, bound by a matrix of white Portland cement (black). An unhydrated cement grain is seen in the centre of view; XPT, ×150.

Portland limestone masonry, which was intended to match the buff colour of the limestone.

Modern Portland cement mortars may incorporate pigments to provide a colour of the purchaser's choice (**318**). They are mainly metallic oxides in the form of a fine, inert powder and are often difficult to disperse evenly throughout a batch of mortar. For this reason pigmented mortar is best obtained as premixed lime:sand coarse stuff or as ready-to-use mortar (Monks, 2000).

## MIX PROPORTIONS

For Portland cement-based mortars and renders, mix proportions are usually determined by simple chemical analysis involving acid dissolution, for example in accordance with the BS 4551 method (British Standards Institution, 2005a) or the ASTM C1342 method (ASTM International, 2005b). These methods give quite good estimations for mixes containing siliceous aggregate. Both standards can cope with Portland cement:sand, Portland cement:lime:sand, and masonry cement:sand mixes. There is the disadvantage that carbonate aggregate particles (limestone, marble, shell) will be dissolved away with the binder giving incorrectly high binder content values. However, estimation of the type and quantity of calcareous aggregate made through the microscope can be used to correct erroneous results of simple chemical analysis.

## WORKMANSHIP

Poor workmanship is probably the biggest cause of early mortar and render failure in new buildings and will, over time, adversely affect the long-term durability of masonry. Batching errors are more likely to occur with site-mixed mortars, where inaccuracies in mix proportions can result from deficiencies in the method used to batch the ingredients. Mortars must be well mixed to achieve their potential strength and this depends on the order in which the ingredients are loaded into the mixer and the mixing time. Symptoms of inadequate mixing include weak/friable mortar and the presence of unmixed lumps of cement (**319**). These symptoms may also indicate the use of stale cement. Over-mixing of mortar will introduce too much air into the mix, in the form of entrapped air voids that will reduce the hardened strength and durability.

Air entrainment can improve the frost resistance of both freshly laid and hardened mortars. BRE Digest 362 (Building Research Establishment, 1991) recommends air contents of between 10% and 18% for this purpose. Figure 320 shows an example of an air-entrained mortar. Too much air entrainment can lead to poor durability, low compressive strength, and poor bond strength (**321**). Excessive air entrainment is a regularly seen defect on

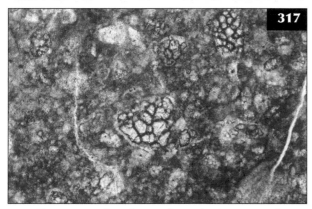

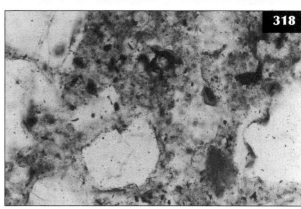

**317** Close view of a white-coloured joint mortar from Portland limestone masonry, comprising white Portland cement mixed with finely crushed Portland limestone particles. Unhydrated white Portland cement grains are frequently seen (white); PPT, ×300.

**318** Close view of a Portland cement-based brickwork mortar coloured yellow by the addition of pigment (possibly synthetic yellow iron oxide). The cement matrix (brown) is seen to enclose a few particles of pigment (orange–brown, lower right). Fine aggregate particles appear white; PPT, ×300.

**319** Unmixed lump of cement (dark brown) within a Portland cement render. The fine cracks (white) seen in the cement lump were probably induced by the oven-drying stage of thin section preparation; PPT, ×35.

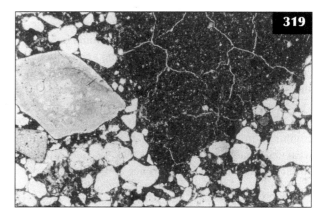

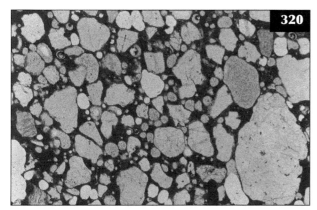

**320** Portland cement:sand mortar exhibiting frequent small, spherical air voids (yellow) indicative of the use of an agent that has imparted a degree of air entrainment (i.e. a chemical admixture). Fine aggregate particles appear white and the cement matrix is shown brown; PPT, ×35.

**321** Weak and friable Portland cement:sand mortar exhibiting abundant small air voids (yellow), following overdosing of the mix with water-reducing admixture (plasticizer). Fine aggregate particles appear white and the cement matrix is shown brown; PPT, ×35.

small building sites, where tradesmen sometimes use household washing-up liquid as plasticizer and accidentally overdose the mix.

Some shrinkage of Portland cement-based mixes is inevitable but this can be controlled by using appropriate ingredients, careful curing, and, if necessary, the provision of movement joints. Drying shrinkage can cause a number of defects including debonding, cracking, and surface crazing.

There are additional workmanship considerations for external Portland cement-based renders. The type of render system used should be carefully selected as being appropriate for the background and exposure conditions in accordance with standard guidance, for example BS 5262 (British Standards Institution, 1991). The background may require special preparation in order to achieve a satisfactory bond. According to current practice, when applying render, individual coats should not be thicker than 16 mm (Building Research Establishment, 1976) as coats thicker than this tend to deform under their own weight and are more prone to drying shrinkage. Each successive coat should be thinner and slightly weaker than the background and/or previous coat, to minimize the risk of shrinkage and resultant shear failure at wall/render or render coat interfaces.

Petrographic examination and chemical analysis for mix proportions and sulfate content are often used in combination to diagnose the reasons for apparent weakness/friability, debonding, cracking, and causes of surface problems like efflorescence and discolouration.

## PHYSICAL AND CHEMICAL DETERIORATION

Masonry exposed to the atmosphere is subjected to natural weathering and will eventually suffer deterioration. Petrographic examination is a fundamental tool used to screen mortars for evidence of decay caused by leaching, salt attack, freeze–thaw damage, and sulfate attack. Microscopical examination will detect signs of matrix replacement, the presence of secondary deposits, and filled or unfilled cracks, which may provide evidence of deterioration and aid diagnosis of its causes.

The presence of secondary deposits indicates that some degree of leaching has taken place, which may have weakened the materials to some degree. As examples of leaching, Figures 322 and 323 show secondary deposits of portlandite and ettringite, respectively, within air voids.

Portland cement mortars are susceptible to sulfate attack, a reaction wherein calcium aluminate phases of the Portland cement are converted to expansive ettringite (calcium aluminium sulfate hydrate) by sulfate-bearing solutions. Figure 324 shows Portland cement render from a chimney, where sulfate attack had caused severe cracking. Under certain conditions Portland cement mortars are potentially susceptible to TSA. This deleterious reaction requires a source of carbonate and often (but not always) cold wet conditions, as well as exposure to sulfates. This results in replacement of cement hydrates by thaumasite (calcium carbonate silicate sulfate hydroxide hydrate). Figure 325 shows an example of TSA in masonry mortar, where thaumasite

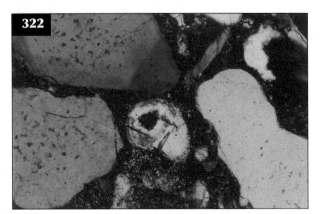

**322** Air void (centre) filled with secondary deposits of portlandite (brightly coloured) by leaching. Portland cement:sand mortar comprising quartz natural sand fine aggregate particles (grey), bound by uncarbonated cement matrix (dark brown); XPT, ×150.

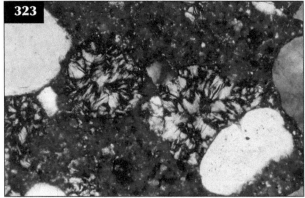

**323** Air voids filled with secondary deposits of ettringite (grey) by leaching. Portland cement:sand mortar comprising quartz natural sand fine aggregate particles (grey/white), bound by uncarbonated cement matrix (dark grey/brown); XPT, ×150.

replacement of the hardened cement matrix had weakened it to the consistency of soft putty.

# SPECIALIST MORTARS

## INTRODUCTION

Specialist mortars have been developed for a variety of particular circumstances including concrete repair, grouting, tiling, mortar jointing, stone filler, sprayed concrete, self-levelling floor finishes, and water-leak stoppers. These tend to be complicated premixed formulations containing multiple ingredients that are manufactured using confidential mix designs. It is often impossible to obtain exact compositional information from manufacturers and the analysis of these materials presents considerable challenges.

The composition of specialist mortars may include any combination of the following ingredients:

- *Cement:* Sometimes more than one cement is used and may include special types of cement. Mixtures of calcium aluminate cement and Portland cement are sometimes used in applications where rapid setting and hardening properties are required, e.g. in repairs to airport runways, marine concrete repair in the tidal zones, and in water-leak sealing. An alternative would be to include rapid hardening cement, which is similar in chemical composition to ordinary Portland cement except that it is more finely ground. For decorative applications such as tiling, mortars may contain white Portland cement

or white calcium aluminate cement to give a white/cream colour.
- *Additions:* Mineral additions of for example, silica fume, GGBS, or PFA.
- *Aggregates:* Normal, lightweight, and special types of aggregates and different types of inert fillers.
- *Admixtures:* Plasticizing chemical admixures are often used. Air entrainers and accelerators are used in some products.
- *Additives:* Many specialist mortars contain a polymer additive to act as a water-reducing agent, act as a bonding agent, increase strength, and reduce permeability. These may include polyvinyl acetate (PVA), polyvinylidene dichloride (PVDC), SBRs, acrylics, and modified acrylics.
- *Fibres:* Concrete repair mortars often contain synthetic fibre reinforcement (usually fine polymer fibres) to help control shrinkage.

When investigating specialist mortars, petrographic examination is typically called upon to help determine the ingredients used and/or investigate the causes of failures.

### PETROGRAPHIC EXAMINATION AND COMPLEMENTARY TECHNIQUES

Petrographic examination of specialist mortars can be performed following guidance given in ASTM C1324 (ASTM International, 2005b), which is largely based on ASTM C856 (ASTM International, 2004). Once a sample is

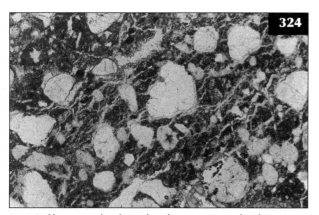

**324** Sulfate attack of Portland cement render from a chimney. Fine/microcracks filled with ettringite run subparallel to the outer surface. Fine aggregate particles appear white and the cement matrix is shown dark brown; PPT, ×75.

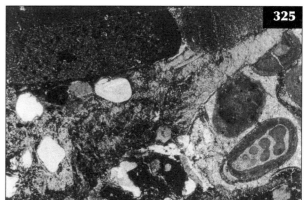

**325** Portland cement-based brickwork mortar suffering from TSA. The mineral thaumasite (yellow) has replaced the cement matrix. Fine aggregate particles include quartz (white), flint (grey), and limestone (brown); XPT, ×35.

received in the laboratory, an initial visual and low-power microscopical examination is conducted to determine the number of material types or layers, layer thickness, colour, and the relative hardness, coherence, and bond strength (in qualitative terms). High-power examination of thin section specimens is then used to determine the grading and mineralogy of any aggregate, binder type, the presence of mineral additions/fillers/ fibres, and to screen the material for evidence of defects or deterioration.

The analysis of specialist mortars presents a substantial challenge owing to the wide range of both inorganic and organic ingredients used. Some ingredients that may comprise less than 1% of the mix and are difficult to detect can have a profound influence on mix properties.

One approach is use separate techniques for the inorganic from the organic ingredients. Optical microscopy is usually most appropriate for the inorganic components, supplemented by scanning electron microscopy if required. Minerals that are difficult to distinguish optically may be readily identified by XRD. Infrared spectrophotometry is the most appropriate method of chemical analysis for identifying organic components. Provided that the correct solvent is used to extract the desired organic species from the sample, this technique can rapidly determine the generic organic materials type and provide clues for the next stage of analysis. Pyrolysis–gas chromatography, gas–liquid chromatography, and ultraviolet spectrophotometry can all be used as complementary or stand alone analytical techniques for isolating a specific organic ingredient (Hunt & Everitt, 2001).

## MIX PROPORTIONS

The standard wet chemical analysis involving acid dissolution, for example in accordance with the BS 4551 method (British Standards Institution, 2005a) or the ASTM C1342 method (ASTM International, 2005b), is unlikely to be accurate for specialist mortars. These methods do not take into account the potentially high contents of mineral additions, calcite fillers, and organic polymers. As a more accurate alternative, mix proportions can be determined petrographically by point-counting the aggregate/cement/ void ratio, either in thin section or polished specimen.

## EXAMPLES OF SPECIALIST MORTARS

The examples below are not intended to be a comprehensive review of mortars for specialist applications. Instead, selected materials have been used to illustrate the range of ingredients that are likely to be encountered in specialist mortars.

Figures 326 and 327 show a proprietary repair mortar for concrete. It consists of fine (1.18 mm nominal maximum sized) quartz fine aggregate bound by Portland cement with polymer modification, a PFA mineral addition, calcite dust filler, and polypropylene microfibres. Figures 328 and 329 show a proprietary mortar designed for setting fastenings in concrete. It consists of 600 µm nominal maximum sized, quartz fine aggregate bound by epoxy resin and Portland cement.

Figure 330 shows a proprietary grey-coloured cementitious tile adhesive. It consists of quartz sand fine aggregate, bound by Portland cement with a calcite dust filler and latex bonding agent. Figures 331 and 332 show a brown-coloured cementitious tile adhesive, comprising a mixture of Portland cement and calcium

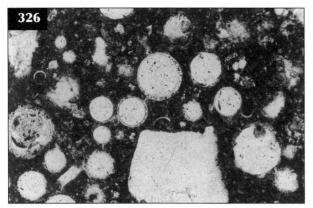

**326** Proprietary repair mortar for concrete, showing fine aggregate particles (white) and cement matrix (dark brown) with PFA (white spheres); PPT, ×150.

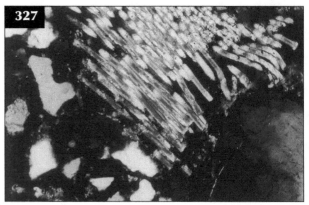

**327** A cluster of polypropylene fibres (brightly coloured) within a proprietary repair mortar for concrete repair; XPT, ×150.

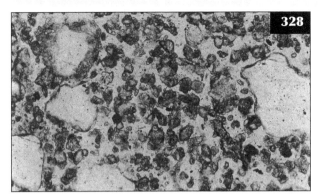

**328** Proprietary mortar for setting fastenings, exhibiting abundant unhydrated Portland cement grains enclosed by epoxy resin. Fine aggregate particles are shown white; PPT, ×150.

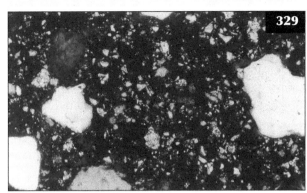

**329** Same view as **328** in cross-polarized light. The epoxy resin in the matrix is isotropic (black), the unhydrated cement grains appear bright, and quartz fine aggregate particles are shown white/grey; XPT, ×150.

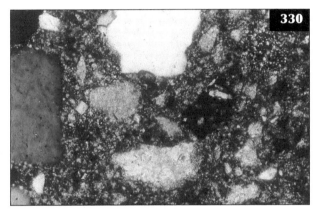

**330** Proprietary grey-coloured cementitious tile adhesive, showing uncarbonated Portland cement binder (brown) with particles of calcite dust filler (light brown). An unhydrated cement grain is seen centre right; XPT, ×300.

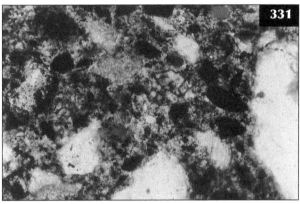

**331** Brown-coloured Portland cement and calcium aluminate cement tile adhesive. Unhydrated/partially hydrated calcium aluminate cement particles are frequently seen (black/red) and an unhydrated Portland cement grain is seen in the centre of view. Quartz fine aggregate is shown white; PPT, ×300.

**332** Same view as **331** in cross-polarized light. Uncarbonated cement matrix appears dark brown and particles of calcite dust filler appear light brown. Quartz fine aggregate particles are shown white; XPT, ×300.

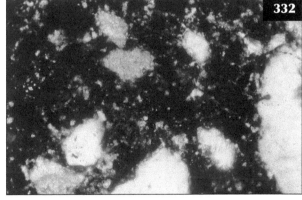

aluminate cement, with calcite filler and quartz fine aggregate. Figures 333 and 334 show a proprietary white-coloured tile adhesive, comprising polymer modified white Portland cement, with calcite filler and quartz fine aggregate. The adhesive is designed to be very fast setting and it includes a latex additive. Figures 335 and 336 show a flexible floor tile adhesive, apparently comprising particles of rubber and calcite dust filler, bound by a matrix of resin and Portland cement.

Figure 337 shows a mortar used as 'stone filler'. A mixture of calcite dust and white Portland cement has been used to fill voids sympathetically that occur naturally in travertine floor tiles.

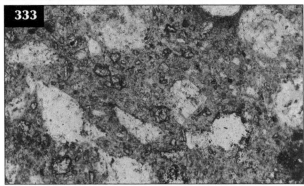

**333** Proprietary white tile adhesive. The cement matrix appears brown and includes a frequent abundance of unhydrated/partially hydrated white Portland cement grains. Fine aggregate particles are shown white; PPT, ×150.

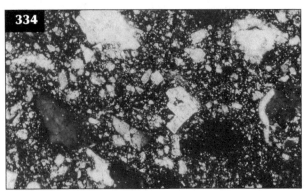

**334** Same view as **333** in cross-polarized light. The uncarbonated polymer modified cement matrix is isotropic (black) and small particles of calcite filler are now clearly seen (light brown). The brightly coloured 'blocky' particle in the centre right is anhydrite; XPT, ×150.

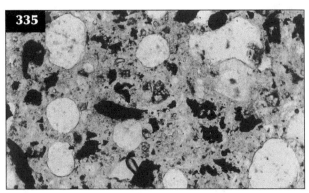

**335** Flexible floor tile adhesive with particles of rubber (black) and calcite dust filler bound by a matrix of resin and Portland cement. Air voids appear yellow and an unhydrated cement grain is seen in the centre of view; PPT, ×150.

**336** Same view as **335** in cross-polarized light. The resin/Portland cement matrix is isotropic (black) and calcite filler particles appear bright; XPT, ×150.

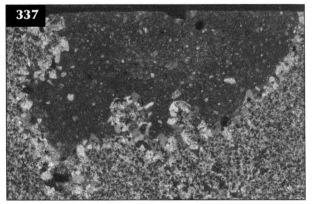

**337** 'Stone filler' mortar (brown) filling a void on the top surface of a natural stone (travertine) floor tile; XPT, ×25. (Courtesy of Mike Eden.)

# Bricks, terracotta, and other ceramics

## INTRODUCTION

The term 'ceramic' is derived from the Greek keramos, meaning 'potter's clay'. Gradually this was extended to include all products made from fired clay and a variety of other nonmetallic inorganic solids. The main types of ceramics used for construction are clay bricks, terracotta, faience, ceramic tiles, and glass.

The earliest bricks used for building were sun-dried mud bricks dating back at least 10,000 years. For stronger brick units, clayey material is fired in a kiln to vitrify, producing a burnt brick. These more resilient fired clay bricks were developed at about 3500BC and by 1200BC fired bricks were being used in durable masonry construction throughout Europe, Asia, and North Africa (Campbell & Pryce, 2003). Clay brick remains an important building material for masonry structures.

Terracotta and faience (glazed terracotta) are moulded clay products comprising hollow blocks, tiles, and sculptural pieces. They are carefully made from specially prepared fine clay mixtures and are usually very well vitrified. They have a hardness and sharpness of detail not normally obtained from brick. They were most widely used in the British and United States building industries in the period 1840–1910AD (the so called 'Terracotta Revival' period).

The use of decorative ceramic tiles for covering walls and floors dates back to ancient Babylon and Egypt and they were later used to great effect by the Persian and Roman civilizations. These traditionally comprised various sizes including small mosaic tiles and were commonly made of a clay-based body, to which a silicate glaze was applied. The familiar large square or rectangular tile appeared around 1600AD as the characteristic blue Delft tile. Tile-making machines began to appear in the early 19th century and today the process is highly automated to meet the demand for large-scale production.

Making glass by heating silica goes back 4000 years to the production of beads and small vessels in Middle Eastern civilisations. Advancements in glass production technology by, amongst others, the Egyptians, the Romans, and the Venetians enabled its use as flat sheets for glazing. Today architectural glass is used for a range of glazing, curtain walling, and tiling applications.

The techniques discussed in the following section can be applied to a much wider range of ceramic materials including pottery, porcelain, and refractories. The principal uses of petrographic examination for investigations of bricks, terracotta, and other ceramics are:

- Identifying the type of material.
- Identifying the ingredients used.
- Providing clues to the details of the manufacturing process.
- Screening the product for inherent defects.
- Identifying the causes of failures.
- Diagnosing decay mechanisms and assessing the level of deterioration.

## PETROGRAPHIC EXAMINATION AND COMPLEMENTARY TECHNIQUES

In the absence of a specific standard procedure for ceramics, petrographic examination procedures could be adapted from the guidance in EN 12407 (British Standards Institution, 2007). An initial visual and low-power microscopical examination is conducted to determine the number of material types, dimensions, manufacturing marks, colour, and the relative hardness, bond strength, and coherence of the material. For examination of smooth glazed surfaces or glass surfaces the use of oblique lighting can be used to pick out surface defects. High-power examination of thin section specimens can then be used to determine the type of ceramic, the nature of the vitrified matrix, the identity of any additives or impurities, and to screen the material for evidence of distress or deterioration. High-power examination of polished specimens is also suitable for investigating ceramics. Polished specimens may be etched to improve the resolution of grain boundaries and other features. A detailed account of the techniques involved in polished specimen preparation, examination, and interpretation for ceramics is provided in Chinn (2003).

Mineralogical analysis by XRD can be used to identify crystalline constituents that cannot be resolved by microscopy. It should be noted that the semiquantitative estimates of composition derived from XRD will be biased due to the presence of amorphous (noncrystalline) phases in ceramics, which cannot be detected by the method. Microstructures and composition of ceramics can also be successfully investigated using SEM and EDS microanalysis (Hernández-Crespo *et al.*, 2007).

# CLAY BRICK

Clay bricks are made by shaping suitable clays or shales to units of standard size, which are then fired at a temperature of 900–1150°C for a period of 8–15 hours. The fired product is a ceramic composed predominantly of silica (55–65%) and alumina (10–25%) combined with as much as 25% of other constituents. Quartz sand is a common and desirable ingredient of brick clay as its presence reduces shrinkage, aids demoulding, and encourages the drying process by creating an open texture. Additional sand is sometimes added to the clay as a 'grog' (or inert filler) and other commonly used grog materials include crushed rock, incinerated domestic refuse, and recycled burnt brick. It is common practice to add some fuel to the brick to produce even firing, for example, coke or powdered coal. Bricks from historic structures can contain all manner of unusual inclusions (Papayianni & Stefanidou, 2003).

Colour is an important characteristic of clay bricks that is dependent on the chemical composition of the raw materials and the nature of the firing process. As a general rule brick colour is determined by the iron content and the calcium carbonate content of the raw materials, combined with the amount of oxygen in the kiln as follows (after Ashurst & Ashurst, 1988):

| Brick colour | Causative constituents and kiln atmosphere |
|---|---|
| Buff | Iron oxide (up to 2%) at 900°C in a reducing atmosphere |
| Bright salmon pink | Iron oxide (up to 2%) at 900°C in an oxidizing atmosphere |
| Red | Iron oxide (up to 2%) at 1100°C in a reducing atmosphere |
| Blue | Iron oxide (up to 7–10%) |
| Black | Iron oxide (up to 7–10%) plus manganese oxides |
| White | Lime (high %) with iron traces |
| Grey | Lime (low %) with iron traces |
| Cream | Chalk (low %) with iron traces |

Colour may not always be uniform and bricks of mixed colour may result from variations in the raw materials, or from thermal gradients developing in the brick during firing. Black patches and black core may result from inclusions of carbonaceous and combustible materials that create reducing conditions within the brick. The superficial colour of the brick may be controlled by treating the outer surface with sand or pigments before firing.

Brick porosity depends both on the clay composition and firing temperature/duration and may range from 1–50%. The vitrified matrix usually contains few micropores (<1 μm across) but there is usually a continuous network of larger voids.

Under favourable circumstances, petrographic examination techniques can be used to identify the source of brick raw materials and the method of manufacturing. The most effective approach is to use a combination of optical microscopy, X-ray diffractometry and SEM with an EDS microanalysis (Pavía, 2006). The origin of clay raw material for bricks is difficult to establish because the raw material is often processed prior to use and then the clay is more or less wholly converted to other minerals by the firing process. If residual clay minerals are present and can be detected and characterized by XRD, it is sometimes possible to trace fired clay products back to a particular clay deposit. When minerals formed at high temperatures are identified the firing temperature in the kiln can be estimated (Dunham, 1992).

Figures **338–340** show historic clay brick from the Great Wall of China (Ming Dynasty, circa 1368–1644AD). An abundance of quartz, feldspar, mica, and granite sand particles are enclosed in the brick. Many of these particles are angular suggesting that crushed granite was deliberately added as 'grog'.

Figures **341** and **342** show historic clay brick from a Jacobean manor house in London (built in 1623). It is relatively compact and includes no large air voids. Quartz and flint sand particles are abundant and these may have originated from the clay raw materials or been added as grog. One flint particle exhibits red discolouration and another appears calcined, suggesting that the brick was heated to between 600°C and 900°C during firing. At higher magnification (**343**) the fired clay matrix has a distinctive red appearance but the detailed structure is still too fine to resolve with the optical microscope. A red clay-rich particle may represent crushed brick grog from a previous firing.

**338** Historic clay brick from the Great Wall of China, comprising particles of quartz (white/grey), feldspar (grey), mica (brown), and granite (large particle) in a matrix of fired clay (black); XPT, ×100.

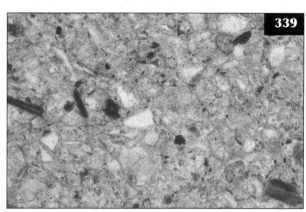

**339** Close view of historic clay brick from the Great Wall of China, showing particles of quartz and feldspar (white) and mica (dark brown) in a fired clay matrix (light brown); PPT, ×100.

**340** Same view as **339** in cross-polarized light. The fired clay matrix now appears black; XPT, ×100.

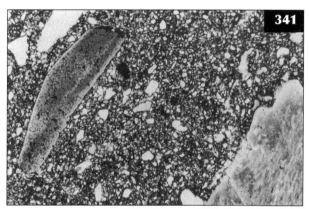

**341** Historic clay brick comprising particles of quartz (white) and flint, with a matrix of fired clay (dark brown). Flint particles exhibit evidence of heating with one exhibiting red discolouration (upper left) and another appearing calcined (lower right); PPT, ×35.

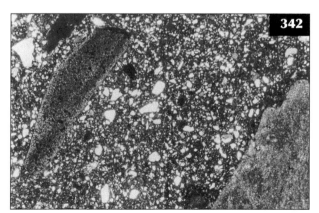

**342** Same view as **341** in cross-polarized light. The fired clay matrix now appears red; XPT, ×35.

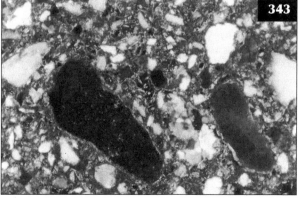

**343** Close view of historic clay brick showing particles of quartz (grey/white) and flint (centre left) with a fired clay matrix (red); XPT, ×150.

Figure **344** shows a modern red clay brick consisting mainly of fired clay matrix and quartz-rich sand particles. The brick is compact with no visible air voids and the clay appears to have been well mixed except for one unmixed lump. Figures **345** and **346** show a modern orange clay brick comprising lumps of fired brick clay that have been pressed together. There is a relatively high content of macropores present as a continuous network of interconnected voids. Figure **347** shows a modern yellow clay brick with uniform texture comprising a matrix of fine, fired clay with almost no sand or other coarse particles. There are some interconnected macropores visible. Figure **348** shows a modern black brick manufactured from Carboniferous shale in England. The brick contains black carbonaceous inclusions originating either from fuel or domestic waste additives.

The main causes of brick masonry decay are eminently suited to diagnosis aided by optical microscopy. The most important deterioration mechanisms are salt crystallization, attack by acid gases in the air, and frost action. Salt crystallization occurs when a solution of salt/s in water is deposited under drying conditions on the surface of the masonry and/or within its pores. The growth of the salt crystals is an expansive process that causes powdering, scaling, and delamination of the outer masonry. Figures **349** and **350** show scaling of the outer surface of brick masonry caused by salt crystallization.

## TERRACOTTA

Terracotta and faience are moulded clay products made from fine, pure clays mixed with 'grog' (or inert filler) to reduce shrinkage considerably. The grog may comprise sand and/or pulverized fired clay. Manufacture involves pressing the carefully prepared clay on to the inside of a plaster mould to a thickness of 1¼ inches (just over 30 mm). The mould is then set aside and the plaster absorbs moisture from the clay. When sufficiently dry the clay is separated from the mould and hand finished as required. After further air drying the clay piece is kiln fired at high temperatures (up to 1250°C). In the case of faience, a high temperature glaze is applied to the unfired clay prior to firing in one operation.

Petrographic techniques used to investigate clay brick can be applied to terracotta to determine the ingredients used, the method of manufacture, and to screen the material for defects and evidence of deterioration. Terracotta units suffer from a similar range of weathering and deterioration mechanisms to those affecting clay bricks. Inadequately fired units are particularly susceptible to frost attack and corrosive atmospheres. In addition, a range of cracking, spalling, and staining issues can arise from corrosion of the iron and steel anchoring systems of terracotta façades.

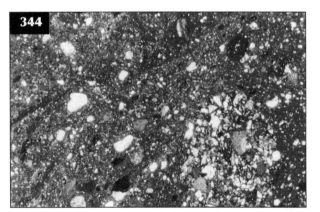

**344** Red clay brick comprising particles of quartz and feldspar (white/grey), with a matrix of fired clay (red). A rounded unmixed lump of fired clay is seen lower right; XPT, ×35.

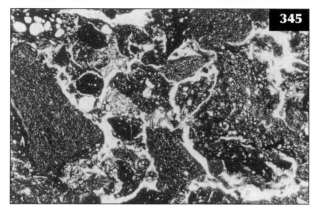

**345** Orange clay brick comprising fired clay (dark brown) with macropores forming a continuous network of interconnected voids (yellow); PPT, ×35.

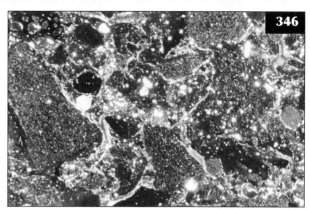

**346** Same view as **345** in cross-polarized light, showing fired brick clay (red) and air voids (dark green); XPT, ×35.

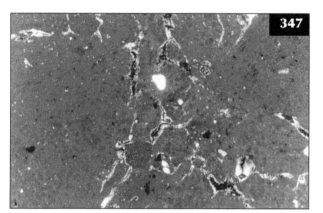

**347** Yellow clay brick comprising a matrix of fired clay (brown) with almost no sand. Interconnected macropores appear dark green; XPT, ×35.

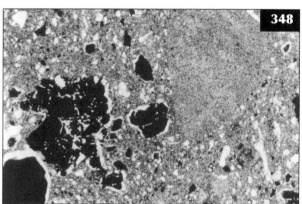

**348** Black brick comprising fired shale (brown) with carbonaceous inclusions (black); PPT, ×35.

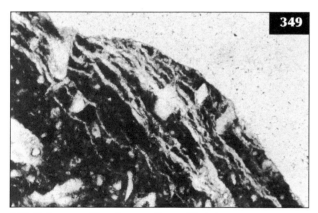

**349** Salt crystallization scaling of brick masonry caused by salt crystallization. Cracks are filled with secondary deposits of gypsum salts (white) while the brick appears dark brown; PPT, ×150.

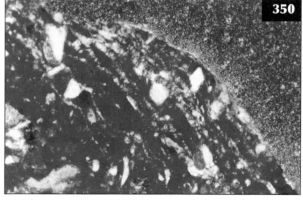

**350** Same view as **349** in cross-polarized light. The fired clay matrix of the brick appears red and the gypsum deposits in the cracks show first-order grey interference colours; XPT, ×150.

Figures 351–353 show terracotta made by the Italian Giovanni da Maiano for Henry VIII in the early 16th century. Manufactured from fine clay with grog comprising a mixture of quartz sand (specially crushed to a uniformly fine size) and pulverized fired clay. XRD detected the presence of the mineral mullite, suggesting a maximum firing temperature for the terracotta in excess of 1000°C.

Figure 354 shows terracotta from the façade of a historic London building, dating from 1905. It is manufactured from fine clay with grog mainly comprising pulverised fired clay. Distinctive wedge-shaped ('arrowheads') crystals of tridymite were observed in thin section suggesting that the brick was heated to temperatures in excess of 870°C during firing (355).

Figure 356 shows an example of modern terracotta which includes a large pellet of previously fired ceramic material as grog.

## CERAMIC TILE

Traditionally ceramic tiles took several days to make as the clay base was fired first and then fired again, after the glaze and any decoration had been applied. Modern methods use a single firing and can be completed in as little as 20 minutes. To achieve a very low and uniform firing shrinkage the bodies of modern tiles are made from finely ground, evenly mixed refractory clays with minimal fluxing agents. A typical recipe would consist of 35% ball clay, 15% china clay, 10% limestone, and up to 15% ground tile waste (Prentice, 1990). The remaining 25% would comprise silica, usually as finely ground quartz sand but frequently including a proportion of cristobalite. Tile glazes are essentially silica, with nepheline syenite or feldspar added as a flux, coloured by metallic oxides, and fused in a furnace. Optical microscopy may be employed to investigate the composition and condition of ceramic tiles and can be complemented by SEM and microanalysis (Kopar & Ducman, 2005). Figures 357 and 358 show a section through the outer surface of a modern ceramic tile. This consists of a fired clay base with an underglaze containing a coloured crystalline aggregate, over which a transparent silica wet glaze has been applied to seal the tile surface.

Ceramic tiles are fixed to walls and floors using tile adhesive (359). The failure of tiled surfaces may involve loss of adhesion or cracking often caused by a lack of provision for movement, use of inappropriate bedding materials, or poor workmanship (Jornet & Romer, 1999). Petrographic examination can provide information helpful to the identification of failure causes, by checking

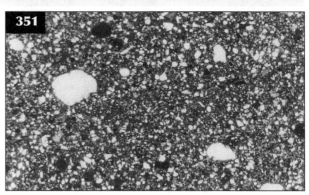

**351** Historic terracotta comprising a fired clay matrix (red) with grog consisting of finely crushed quartz-rich sand (white); XPT, ×35.

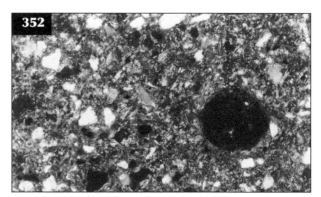

**352** Close view of historic terracotta showing fired clay matrix (red/brown) and grog particles consisting of angular quartz particles (grey/white) and previously fired clay (red, right); XPT, ×150.

**353** Very close view of historic terracotta showing fired clay matrix (red) and grog particles consisting of angular quartz particles (grey/white) and muscovite mica (pink, centre); XPT, ×300.

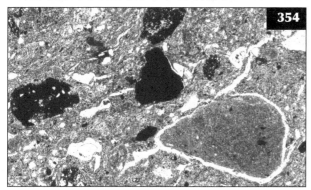

**354** Historic English terracotta showing a matrix of fired brick clay with grog comprising particles of previously fired clay (rounded particles); PPT, ×35. (Courtesy of Barry J Hunt.)

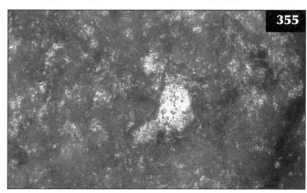

**355** Close view of English terracotta manufactured in 1905 showing an arrowhead-shaped crystal of tridymite (grey); XPT, ×300. (Courtesy of Barry J Hunt.)

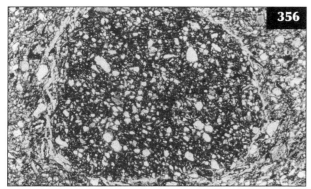

**356** A large ceramic grog pellet in terracotta; PPT, ×35.

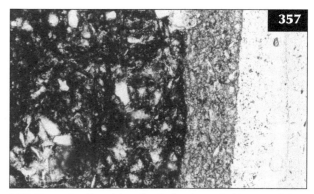

**357** Ceramic tile with a fired clay base (left) comprising vitified clay (brown) with angular inclusions of ground quartz (white). Opaque underglaze applied to clay base (grey) includes fine particles of a crushed aggregate for colour. The outer glaze (right) is glassy and transparent (white); PPT, ×150.

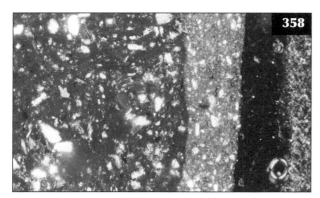

**358** Same view as **357** in cross-polarized light. The fired clay base appears red with grey/white quartz inclusions (left). The underglaze containing fine-grained colouring aggregate appears grey and the glassy outer glaze is isotropic (black, right); XPT, ×35.

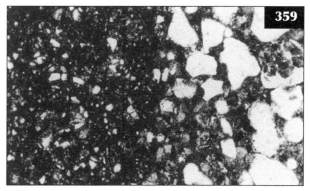

**359** Ceramic floor tile (left) and tile adhesive (right). The tile consists of fired-clay matrix (reddish-brown) with very fine quartz sand inclusions (white). The adhesive comprises natural sand fine aggregate (white) bound by a cement matrix consisting of HAC mixed with Portland cement (brown); PPT, ×75.

the composition and condition of the bedding/jointing materials and in the assessment of crack morphologies (Wong *et al.*, 2003).

## ARCHITECTURAL GLASS

Modern flat glass for architectural applications is commonly manufactured by the float glass process, or less frequently using the older sheet process, or the rolled process. It is a soda–lime–silica glass with a typical raw material composition (by weight) of silica sand (72%), soda ash (13%), limestone (10%), and dolomite (4%). Coloured glasses are produced by additions of small amounts of colouring agents such as iron (green), nickel (brown) or cobalt (blue).

Microscopical examination of glass can provide much useful information. It affords a quick means of identification of glass type through measurement of the refractive index. Glass comprises amorphous silica that appears isotropic under the optical microscope. Although relatively uniform in composition it may contain impurities and imperfections, of which the frequency, size, and sources may be determined.

Imperfections include bubbles (or 'seeds') that may have a number of possible sources, the most common being gas evolved during firing. Bubbles may contain crystalline materials formed during cooling of the glass that may provide clues to the origin of the bubbles. Cords are linear features within the glass that may result either from imperfectly homogenized raw materials, dissolved refractories or devitrified material. Figure 360 shows the appearance of soda–lime–silica glass that exhibits bubbles and cords. 'Stones' are solid crystalline substances occurring in glass that are regarded as defects. They are usually derived either from the batch material, refractories, or devitrification. Figure 361 shows the appearance of soda–lime–silica glass that contains a devitrification 'stone'. These may develop as the result of incomplete mixing of the molten glass constituents and/or too low a firing temperature. The 'stone' shown in Figure 361 contains an aggregation of tridymite crystals (see 362).

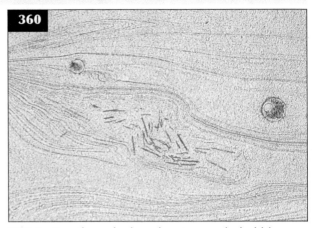

**360** Section through glass showing cords, bubbles, and a devitrification 'stone' (centre); PPT, ×100. (Courtesy of Barry J Hunt.)

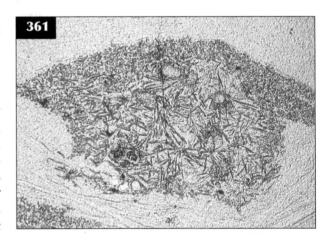

**361** Section through glass showing a devitrification 'stone' associated with a fine crack; PPT, ×150. (Courtesy of Barry J Hunt.)

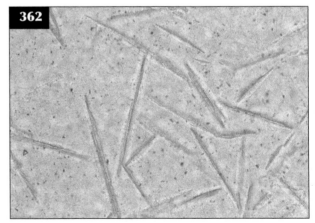

**362** Close view of tridymite plates in glass; PPT, ×300. (Courtesy of Barry J Hunt.)

# Bituminous mixtures

## INTRODUCTION

The term 'bituminous mixtures' is used to denote all materials in which an aggregate is bound with a hydrocarbon binder. They are most extensively used as surfacing material for road construction, where they are commonly referred to as 'blacktop'. They were first produced in the 1870s and became extensively available by the 1930s.

Roads are built from a number of layers that together are called a pavement. Bituminous mixtures are used in the upper layers of flexible or flexible-composite pavements, as opposed to rigid pavements where the top layers comprise concrete. Typical flexible pavement structure is illustrated in Figure 363.

Bituminous mixtures contain two main ingredients, which are aggregate (coarse and fine) and the binder (petroleum bitumen, natural bitumen, road tar). Various grades of each may be used in various proportions to produce mixes with different design properties. Fillers (fine-grained material such as limestone dust, Portland cement or PFA) may also be included in the mix to change its physical characteristics.

Various bituminous mixtures have evolved to suit the range of different circumstances in which they are used. They can broadly be divided into two groups, the macadams and asphalt. Macadams have a high content of well graded (dense graded) aggregate and a low content of bitumen binder, giving a dense stable structure with the load being transmitted through the aggregate. In

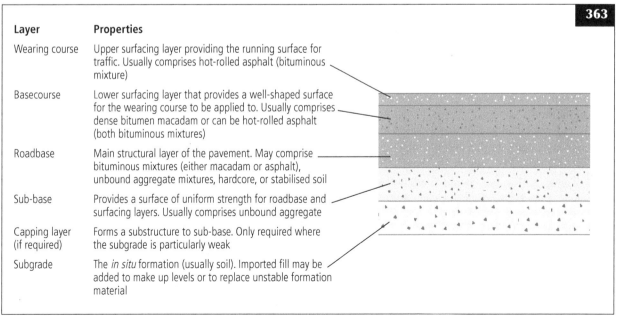

| Layer | Properties |
|---|---|
| Wearing course | Upper surfacing layer providing the running surface for traffic. Usually comprises hot-rolled asphalt (bituminous mixture) |
| Basecourse | Lower surfacing layer that provides a well-shaped surface for the wearing course to be applied to. Usually comprises dense bitumen macadam or can be hot-rolled asphalt (both bituminous mixtures) |
| Roadbase | Main structural layer of the pavement. May comprise bituminous mixtures (either macadam or asphalt), unbound aggregate mixtures, hardcore, or stabilised soil |
| Sub-base | Provides a surface of uniform strength for roadbase and surfacing layers. Usually comprises unbound aggregate |
| Capping layer (if required) | Forms a substructure to sub-base. Only required where the subgrade is particularly weak |
| Subgrade | The *in situ* formation (usually soil). Imported fill may be added to make up levels or to replace unstable formation material |

**363** Typical flexible pavement structure.

contrast, asphalts have a low content of single-sized (open graded) aggregate, with a high bitumen content and a high filler/fines content. These provide a dense, strong, and stiff layer with the load being transmitted through the binder. Due to the low aggregate content, bitumen-coated stone chippings are often rolled into the upper (trafficked) surface of asphalt wearing courses to increase the skid resistance.

Other bituminous mixtures that the petrographer should be aware of include mastic asphalt, sand carpet, porous asphalt, stone mastic asphalt (SMA), and grouted macadam. Mastic asphalt is dense, voidless asphalt that is rich in hard bitumen and filler. The filler used is coarser than that used in road asphalt. The mastic may be extended with coarse aggregate depending on end-use. It is impermeable and used to surface large flat roofs, multistorey car park decks, and bridge decks. Sand carpet is similar to mastic asphalt in that it is bitumen rich, but contains less filler, the remainder of the mix being made up with fine aggregate. Unlike mastic asphalt, it contains no coarse aggregate. It is used in several specialized applications such as on bridge decks where it is used to protect underlying waterproofing polymer membranes (sand carpet is sometimes coloured red when used for this application).

Porous asphalt comprises uniformly graded (open-graded) aggregate with low bitumen content and an open texture (20% large interconnected air voids). It is free draining and is used on airfield pavements to prevent aquaplaning and it is increasingly used on roads as it reduces the rolling noise of traffic.

Stone mastic asphalt (SMA) comprises gap-graded aggregate with voids almost completely filled by a rich mastic of bitumen/fine aggregate/filler. Cellulose fibres or polymer modified bitumen may also be incorporated. The material was designed in Germany to resist studded snow tyres and SMA is now being used in the UK and North America as it has good resistance to rutting and high durability.

Grouted macadam is laid using a two-stage process and comprises open-graded asphalt topping with a void content of 20–25%, which is later flooded with cementitious mortar. The resulting composite has a high resistance to point loading and chemical/fuel spillage. It is being increasingly used at container parks, docks, airport aprons, and garage forecourts.

The common causes of premature deterioration in bituminous mixture surfacings are related to loss of structural strength and wear, disintegration, or loss of surface characteristics. Currently under-utilized, microscopy can provide valuable information regarding the causes and magnitude of road construction material deterioration.

The principal applications of petrographic examination to investigation of bituminous mixtures are:
- Determining the number material types and layers present (and their thickness).
- Identifying aggregates and filler used.
- Determination of the air void structure.
- Identifying the causes and extent of defects and deterioration.

## PETROGRAPHIC EXAMINATION

In the absence of a specific standard procedure for petrographic examination of bituminous mixtures, the guidance of EN 12407 (British Standards Institution, 2007) or ASTM C856 (ASTM International, 2004) may be adapted. An initial visual and low-power microscopical examination is conducted to determine the number of layers, their thickness, and to look for any macroscopic evidence of deterioration. Coarse aggregate and chippings exposed at the upper surface of roads can be examined for evidence of damage. Observation of slices cut perpendicular to the road surface is particularly useful for studying the properties of coarse aggregate. Details of the aggregate size, shape, and distribution can be accurately determined using image analysis techniques (Schlangen, 1999). Slices that have been impregnated with fluorescent resin are useful for observing the air void structure (Eriksen, 1999). High-power microscopical examination of thin sections is used to identify the surface texture, geological type, and potential durability of aggregates and fillers used and to assess the effectiveness of the bitumen coating.

## EXAMPLES OF BITUMINOUS MIXTURES

Figures **364** and **365** show macadam from a roadbase, comprising well graded crushed dacite aggregate with a low content of bitumen binder that barely coats the aggregate and leaves air voids between aggregate particles. Figure **366** shows hot-rolled asphalt wearing course comprising single-sized crushed rock aggregate with a relatively high bitumen content, leaving no voids. Calcium carbonate dust filler is clearly seen within the

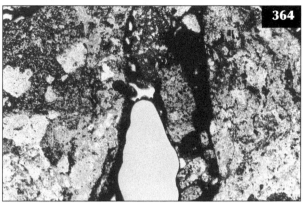

**364** Macadam with crushed dacite aggregate (pale green) and bitumen binder (black). Air void shown yellow; PPT, ×35.

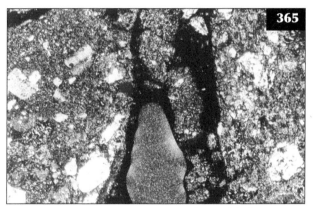

**365** Same view as **364** in cross-polarized light. The dacite aggregate appears grey, bitumen black, and an air void is shown dark green; XPT, ×35.

**366** Close view of hot-rolled asphalt with crushed volcanic rock coarse aggregate particles (grey), bitumen binder (dark brown), and calcium carbonate filler (pink, encapsulated by the bitumen); XPT, ×150.

binder. Figure **367** shows sand carpet from a bridge deck. It consists of natural sand fine aggregate (with no coarse aggregate) bound by a relatively high content of bitumen. Figures **368** and **369** show a grouted macadam from a car park, consisting of open textured asphalt with the voids filled by cementitious mortar. The cementitious mortar comprises 25% of the material and contains an addition of microsilica. Figure **370** shows a bituminous bedding material of an historic wood block floor (circa 1810). The material comprises a mixture rich in bitumen (possibly a natural bitumen) combined with a natural sand fine aggregate.

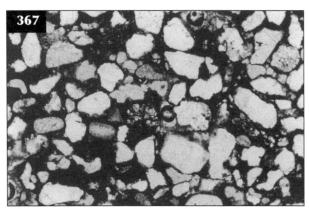

**367** Sand carpet comprising bitumen (black) mixed with natural sand fine aggregate (white). Small air voids are shown yellow; PPT, ×150.

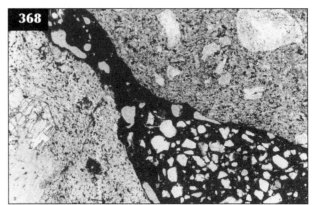

**368** Grouted macadam with crushed rock coarse aggregate (intermediate igneous rock, white/light brown), coated by bitumen (black). Large voids filled by microsilica improved sand:cement mortar with the sand particles appearing white (lower right). Small voids are unfilled and shown yellow; PPT, ×35.

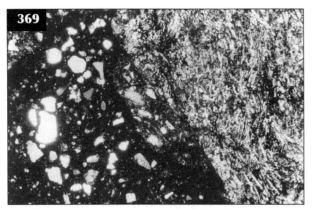

**369** Grouted macadam with crushed rock coarse aggregate (intermediate igneous rock, grey), coated by bitumen (brown, centre). Large voids are filled by microsilica improved sand:cement mortar (black, lower left) with the sand particles appearing white; XPT, ×35.

**370** Historic bituminous floor bedding, with natural sand consisting of quartz (grey/white) and phosphatic rock particles (brown), bound by bitumen (black); XPT, ×35.

# Appendix A

## SUPPLIERS OF PETROGRAPHIC EQUIPMENT, TRAINING, AND LITERATURE

Many of the companies listed below have regional offices and foreign distributors that are listed on their web sites.

### COMPANY/PRODUCTS

Leica Microsystems GmbH
Ernst-Leitz-Strasse 17-37,
35578 Wetzlar, Germany
Telephone: +49 6441 29 0
Website: www.leica-microsystems.com
*High quality microscopes and associated imaging systems, including the Leica DM range of polarizing microscopes.*

Carl Zeiss AG
Carl-Zeiss-Strasse,
D-73447 Oberkochen, Germany
Telephone: +49 7364 20 6175
Website: www.zeiss.de
*High quality microscopes and associated imaging systems, including the Zeiss Axio Imager polariszing microscope.*

Olympus Corporation
Shinjuku Monolith,
3-1 Nishi-Shinjuku 2-chome,
Shinjuku-ku, Tokyo 163-0914, Japan
Website: www.olympus.co.jp
*High quality microscopes and associated imaging systems, including the Olympus BX range of microscopes.*

Nikon Corporation
Fuji Bldg., 2-3, Marunouchi 3-chome, Chiyoda-ku, Tokyo 100-8331, Japan
Telephone: +81 3 3214 5311
Website: www.nikon.com
*High quality microscopes and associated imaging systems, including the Nikon Eclipse series of polarizing microscopes.*

Logitech Ltd
Erskine Ferry Road, Old Kilpatrick,
Glasgow, G60 5EU, UK
Telephone: +44 1389 875444
Website: www.logitech.uk.com
*Thin section and polished specimen preparation machines and consumables, including the LP50 auto-lap lapping machine for thin sections.*

Buehler Ltd
41 Waukegan Rd, P.O. Box 1
Lake Bluff, IL 60044-1699, USA
Telephone: +1 847 295 6500
Website: www.buehler.com
*Thin section and polished specimen preparation machines and consumables, including the PetroThin thin section making system.*

German Instruments A/S
Emdrupvej 102
DK 2400 Copenhagen NV, Denmark
Telephone: +45 39 67 71 17
Website: www.germann.org
*A thin section preparation machine using diamond grinding rollers.*

Hillquist, Inc
1144 S. Bannock Street
Denver, CO 80223, USA
Telephone: +1 303 722 9766
Website: www.hillquist.com
*Thin section and polished specimen preparation machines and consumables.*

Struers A/S
Pederstrupvej 84, 2750 Ballerup, Denmark
Telephone: +45 44 600 800
Website: www.struers.com
*Thin section and polished specimen preparation machines and consumables, including the Discoplan T/S thin section making machine.*

Microtec Engineering Laboratories, Inc
P.O. Box 636
Clifton, CO 81520, USA
Telephone: +1 970 434 8883
Website: www.microteclab.com
*Thin section preparation machines and consumables, including the Micro-Trim automatic thin-sectioning machine.*

Conwy Valley Systems Ltd
Wells House, 12, Hawes Drive, Deganwy,
Conwy, LL31 9BW, UK
Website: www.steppingstage.com
*Mechanical stepping stage that fits to the microscope stage to enable point-counting of material proportions.*

Concrete Experts International ApS
Marielundvej 30, 1. sal, DK-2730 Herlev Denmark
Telephone: +45 3916 1660
Website: www.concrete-experts.com
*RapidAir 457: An automated machine for analysis of concrete air void systems in accordance with ASTM C457 or EN 480-11. W/C Check: Computer software for automatic determination of water/cement ratio in specimens of hardened concrete impregnated with fluorescent resin.*

Oxford Cryosystems Ltd
3 Blenheim Office Park
Lower Road, Long Hanborough
Oxford OX29 8LN, UK
Telephone: +44 1993 883488
Website: www.oxfordcryosystems.co.uk
*Metripol birefringence imaging microscope unit.*

British Standards Institute (BSI)
British Standards House
389 Chiswick High Road
London W4 4AL, UK
Telephone: 0208 996 9000
Website: www.bsi-global.com
*British and European standard procedures and literature.*

ASTM International
100 Barr Harbor Drive, West Conshohocken, Pennsylvania, USA
Telephone: +1 610 832 9500
Website: www.astm.org
*American standard procedures and literature.*

# Appendix B

## STAINING TECHNIQUES FOR GEOMATERIALS PETROGRAPHY

| Subject | Method summary* | Observation |
|---|---|---|
| Carbonate minerals | Dickson's method[1,2] is the most useful scheme for *carbonates* which in one operation differentiates between nonferroan and ferroan phases in both calcites and dolomites: etch for 10 s in 1.5% HCl. 30 s immersion in 1:1 g Alizarin Red S + 0.9 g potassium ferricyanide in 100 ml 0.5% HCl. 10 s immersion in 0.2 g Alizarin Red S in 100 ml 1.5% HCl, wash in water. | Calcite stains pink to red. Ferroan calcite stains purple (mauve) to dark blue. Dolomite is unstained. Ferroan dolomite stains very light blue (turquoise). Aragonite stains pink to red. |
| | *Dolomite* is often recognized by its failure to stain when using Dickson's method. For specific dolomite detection use a method such as this one described by Tucker[3]: 5 min immersion in a boiling solution of Titan Yellow solution (0.2 g Titan Yellow dissolved in 25 ml methanol by heating, top up to 25 ml with more methanol and then add 15 ml of 30% sodium hydroxide solution). | Dolomite stains deep orange red. Magnesite stains orange to red. High magnesium calcite stains orange. Calcite and aragonite remain unstained. |
| | Haines[4] method detects *brucite* and dolomite: etch for 3–5 s in 5% HCl. Apply reagent (0.2% Alizarin Red S in 0.2% HCl) drop by drop until strong pink colour develops on calcite, wash in water. | Brucite stains purple and calcite pink, while dolomite remains unstained. |
| | Fiegl[5] method specifically for *aragonite*: 10 min immersion in Fiegl's solution at 20°C. Fiegl's solution is made by adding 1g silver sulfate to a boiling solution of 11.8 g manganese sulfate in 100 ml of water. Cool and filter the suspension; add 2 drops 10% sodium hydroxide solution. Filter into a dark storage bottle after 2 h. | Aragonite is stained black while calcite and dolomite remain unstained. |
| Feldspars | Houghton[6] method for staining plagioclase and alkali feldspars: etch over hydrofluoric acid vapour (55% HF solution) for 25–35 s. 45 s immersion in saturated sodium cobaltnitrate solution. Gently wash in water, blot dry, and then 2 s immersion in 5% barium chloride solution. Water wash and place several drops of potassium rhodizonate solution on surface (0.01 g K-rhodizonate in 30 ml distilled water). Leave until plagioclase grains become pink and wash in water. | Alkali feldspars stain yellow. Pure na-albite remains unstained; other plagioclases stain pale to deep pink in proportion to the amount of calcium in the molecule: albite/oligoclase will stain lighter than a more calcic plagioclase. Quartz remains unstained. |
| Feldspathoids | Shand[7] method for feldspathoids: spread a thin layer of 85% (syrupy) phosphoric acid over specimen with glass rod. Wait 3 min and wash gently in water. Immerse in 0.25% methylene blue solution for 1 min, wash in water. | Nepheline, sodalite, and alacime stain deep blue. Melilite stains pale blue. Leucite remains unstained. Zeolites also stain blue. |

| Subject | Method summary* | Observation |
|---|---|---|
| Clay minerals | Tucker[3] describes a simple technique to visualize the distribution of clays in sandstones and argillaceous limestone: soak grease-free specimen surfaces in solution of 0.5 g malachite green, congo red, or methylene blue in 250 ml of water (choose the dye colour that contrasts most with the rock). Also, replacing half of the water with ethanol or methanol works well on limestones, including chalk. | Clay minerals take up the stain. |
| Sulfate minerals | Poole and Thomas[8] method for *sulfate minerals*: 2 min immersion in 2:1 mixture of $BaCl_2$:$KMnO4$ 6% solution, wash first with water then saturated oxalic acid. | Gypsum, anhydrite, ettringite, and gypsum plasters stain pink to purple. |
| | Friedman[9] method for *sulfate minerals*: a few minutes' immersion in a cold solution of 0.1–0.2 g Alizarin Red S in 25 ml of methanol added to 50 ml of 5% sodium hydroxide. | Gypsum stains purple, anhydrite and calcite remain unstained, while dolomite stains very pale purple. |
| | Hounslow[10] method for *gypsum* and *anhydrite*: a few seconds' immersion in a 10% solution of mercuric nitrate acidified by 1% nitric acid. | Gypsum and anhydrite stain yellow. |
| Alkali–silica gel | Natesaiyer and Hover[11,12] method: 15 min immersion in 19% uranyl acetate 1.5% acetic acid solution, wash in water. | Ultraviolet light at 240 nm wavelength gives alkali–silica gel a green fluorescence. |
| | Poole, McLachlan, and Ellis[13] method: immerse in 4M cuprammonium sultate for 72 h and then wash in water. | Alkali–silica gel in concrete voids and cracks stain blue. |
| | Guthrie and Carey method[14,15]: immerse in saturated solution of sodium cobaltnitrate, followed by immersion in Rhodamine B base solution. | Alkali-rich alkali–silica gel stains bright yellow. Calcium-rich gel stains pink. |

* Full method details and safety requirements contained in the original references (listed below) should be consulted before a method is used for the first time.

1   Dickson JAD (1965). A modified staining technique for carbonates in thin section. *Nature* 205:587.
2   Dickson JAD (1966). Carbonate identification and genesis as reveled by staining. *Journal of Sedimentary Petrology* 361/2:491–505.
3   Tucker M (ed) (1988). *Techniques in Sedimentology*. Blackwell Science Ltd, London.
4   Haines M (1968). Two staining tests for brucite in marble. *Mineralogical Magazine* 36:886–888.
5   Fiegl F (1937). *Qualitative Analysis by Spot Tests*. Nordemann, New York.
6   Houghton HF (1980). Refined techniques for staining plagioclase and alkali feldspars in thin section. *Journal of Sedimentary Petrology* 50:629–631.
7   Shand SJ (1939). On staining of feldspathoids and on zonal structure of nepheline. *American Mineralogist* 24:508–513.
8   Poole AB, Thomas A (1975). A staining technique for the identification of sulfates in aggregates and concrete. *Mineralogical Magazine* 40:315–316.
9   Friedman GM (1959). Identification of carbonate minerals by staining methods. *Journal of Sedimentary Petrology* 29:87–97.
10  Hounslow AW (1979). Modified gypsum/anhydrite stain. *Journal of Sedimentary Petrology* 49:636–637.
11  Natesaiyer KC, Hover KC (1988). *In situ* identification of ASR products in concrete. *Cement and Concrete Research* 18:455–463.
12  Natesaiyer KC, Hover KC (1989). Further study of *in situ* identification method for alkali-silica reaction products in concrete. *Cement and Concrete Research* 19:770–778.
13  Poole AB, McLachlan A, Ellis DJ (1988). A simple staining technique for the identification of alkali-silica gel in concrete and aggregate. *Cement and Concrete Research* 18:116–120.
14  Guthrie GD, Carey JW (1997). A simple environmentally friendly and chemically specific method for the identification and evaluation of alkali-silica gel. *Cement and Concrete Research* 27/9:1407–1417.
15  Guthrie GD, Carey JW (1998). *A Geochemical Method for the Identification of ASR Gel*. TRB Paper No. 991261. Transportation Research Board, Washington, USA.

## ETCHING AND STAINING TECHNIQUES FOR CEMENT MINERALS AND SLAGS

| Subject | Method summary* | Observation |
|---|---|---|
| Portland cement | *Aluminates* | |
| | **Hydrofluoric acid vapour**[1]: hold a few millimetres above acid and etch in vapour for 2–10 s at room temperature. | Tricalcium aluminate ($C_3A$) turns grey. Tetracalcium aluminoferrite ($C_4AF$) turns white. Also stains silicates (see below). |
| | **5% potassium hydroxide in ethyl alcohol solution**[1]: immerse for 20 s, wash with 1:1 ethyl alcohol, water, then isopropyl alcohol buff (with microlap cloth) for 15 s then wash with isopropyl alcohol. | Tricalcium aluminate ($C_3A$) turns blue. |
| | **Warm (40°C) distilled water**[1]: immerse 5–10 s. | Aluminates turn blue to brown. Tricalcium silicate (alite, $C_3S$) light tan. Dicalcium silicate (belite, $C_2S$) is unchanged. Calcium hydroxide is multicoloured. |
| | *Silicates* | |
| | **Dilute salicylic acid stain**[1]: solution comprises 0.2 g salicylic acid + 25 ml isopropyl alcohol + 25 ml water. Immerse for 20–30 s. | Tricalcium silicate (alite, $C_3S$) and dicalcium silicate (belite, $C_2S$) turn blue–green. Alite stains faster so may be deeper coloured. |
| | **Salicylic acid etchant**[1]: solution comprises 0.5 g salicylic acid + 50 ml methyl alcohol. Immerse for 45 s. | Alite and belite visibly etch. Belite shows lamellar structure. |
| | **Nital**[1]: nital solution comprises 1.5 ml nitric acid + 10 ml isopropyl alcohol. Immerse for 6–10 s. | Alite turns blue to green. Belite turns brown to blue. |
| | **Hydrofluoric acid vapour**[1]: hold a few millimetres above acid and etch in vapour for 5–10 s at room temperature. | Alite turns brown and belite turns blue. |
| | *Hydrates* | |
| | **Napththol green B**[2]: immerse thin section for 12 h in an alcoholic solution of the dye containing a few percent water. | Calcium hydroxide turns green. |
| High-alumina cement | *Aluminates* | |
| | **Boiling 10% sodium hydroxide solution**[1]: immerse for 20 s. | Monocalcium aluminate (CA) turns blue or brown. |
| | **1% borax solution**[3]: immerse for 30 s. | Dodeca-calcium hepta-aluminate ($C_{12}A_7$) turns grey. |
| | **Boiling 1% borax solution**[3]: immerse for unknown time. | Pleochroite (approximately $C_{22}A_{13}F_3S_4$) is etched. |
| | *Silicates* | |
| | As for Portland cement, although HAC has no tricalcium silicate but does contain limited amounts of dicalcium silicate. | |

| Subject | Method summary* | Observation |
|---|---|---|
| Slags | *GGBS* | |
| | **Hydrofluoric acid vapour** [4]: hold a few millimetres above acid and etch in vapour for 2–10 s at room temperature. | GGBS particles turn white to blue. |
| Blended cement | *Blastfurnace cement* | |
| | **1% nitric acid solution**: immerse for 5–20 s (useful for estimating the slag content of blended cement composed of Portland cement and GGBS). | GGBS unchanged. Silicates turn blue, aluminates and ferrite golden. |

*  Full method details and safety requirements contained in the original references (listed below) should be consulted before a method is used for the first time.

1   Campbell DH (1999). *Microscopical Examination and Interpretation of Portland Cement and Clinker*, 2nd edn. Construction Technology Laboratories, a Division of the Portland Cement Association, Skokie, Illinois.
2   Lea FM (1970). *The Chemistry of Cement and Concrete*, 3rd edn. Edward Arnold (Publishers) Ltd, London.
3   Long GR (1983). Microstructure and chemistry of unhydrated cements. *Philosophical Transactions*, Royal Society of London, A310, 42–51.
4   St John DA, Poole AB, Sims I (1997). *Concrete Petrography, a Handbook of Investigative Techniques*. Edward Arnold (Publishers) Ltd, London.

# References and further reading

## CHAPTER 1

Allman M, Lawrence DF (1972). *Geological Laboratory Techniques*. Blandford Press, Poole, Dorset.

ASTM International (2008). *Standard Guide for the Petrographic Examination of Aggregates for Concrete*. ASTM C295-08. Philadelphia, USA.

Bradbury S, Bracegirdle B (1998). *Introduction to Light Microscopy*. Royal Microscopical Society Microscopy Handbook 42. Bios Scientific Publishers, Oxford.

British Standards Institution (1997). *Tests for General Properties of Aggregates – Part 3: Procedure and Terminology for Simplified Petrographic Description*. BS EN 932-3. BSI, London.

Campbell DH (1999). *Microscopical Examination and Interpretation of Portland Cement and Clinker*, 2nd edn. Construction Technology Laboratories, a Division of the Portland Cement Association, Skokie, Illinois.

Davey N (1961). *A History of Building Materials*. Phoenix House, London.

Doran DK (ed) (1992). *Construction Materials Reference Book*. Butterworth-Heinemann Ltd, Oxford.

Doran DK, Cockerton CR (2006). *Principles and Practice of Testing in Construction. Testing in Construction Volume 1*. Whittles Publishing Limited, Scotland.

Entwistle A (2002). Making and displaying your own microscopy web page. *Proceedings of the Royal Microscopical Society* 37/4:224–228.

Entwistle A (2003a). Creating animated digital micrographs: making movies. *Proceedings of the Royal Microscopical Society* 38/2:74–78.

Entwistle A (2003b). Basic digital light micrography. *Proceedings of the Royal Microscopical Society* 38/4:236–245.

Entwistle A (2004). Basic digital light micrography, part 2. *Proceedings of the Royal Microscopical Society* 39/1:15–24.

Fookes PG (1991). Geomaterials. *Quarterly Journal of Engineering Geology* 24:3–15.

Geological Society of America (1991). *Rock-Color Chart with Genuine Munsell Color Chips*. Geological Society of America, Boulder, Colorado.

Humphries DW (1992). *The Preparation of Thin Sections of Rocks, Minerals and Ceramics*. Royal Microscopical Society Microscopy Handbooks. Oxford Science Publications, Oxford.

Hutchison CS (1974). *Laboratory Handbook of Petrographic Techniques*. Wiley Interscience, New York.

Jana D (2006). Sample preparation techniques in petrographic examinations of construction materials: a state-of-the-art review. *Proceedings of the 28th Conference on Cement Microscopy*, ICMA, Denver, Colorado, 23–70.

Kerr PF (1977). *Optical Mineralogy*. McGraw-Hill Books, New York.

McNally GH (1998). *Soil and Rock Construction Materials*. E & FN Spon, London.

Munsell Soil Color Charts (1994). *Revised Edition Soil Color Charts*. Macbeth Division of Kallmorgan Instruments Corporation, New Windsor.

Murphy CP (1986). *Thin Section Preparation of Soils and Sediments*. AB Academic Publishers, Berkhamsted.

Prentice JE (1990). *Geology of Construction Materials*. Chapman and Hall, London.

Russ JC (1990). *Computer-assisted Microscopy: Measurement and Analysis of Images*. Kluwer Academic/Plenum Publishers, New York.

Russ JC (2006). *The Image Processing Handbook*. CRC Press Inc, Boca Raton.

Taylor LE, Brown TJ, Lusty PAJ, *et al.* (2006). *United Kingdom Minerals Yearbook 2005*. British Geological Survey. Keyworth, Nottingham.

Woodcock NH (1994). *Geology and Environment in Britain and Ireland*. UCL Press, London.

## CHAPTER 2

Adams AE, Mackenzie WS, Guildford C (1984). *Atlas of Sedimentary Rocks under the Microscope*. Longman Group Limited, Harlow.

Alm D, Brix S, Howe-Rasmussen H, Hansen KK, Grelk B (2005). Etching and image analysis of the microstructure in marble. *Proceedings of the 10th Euroseminar on Microscopy Applied to Building Materials*. Paisley, United Kingdom, 21–25 June 2005.

Arnold D (1991). *Building in Egypt: Pharaonic Stone Masonry*. Oxford University Press, Oxford.

Ashurst J, Dimes FG (1998). *Conservation of Building and Decorative Stone*. Butterworth Heinemann, Oxford.

ASTM International (2003). *Standard Specification for Granite Dimension Stone*. ASTM C615-03. Philadelphia, USA.

ASTM International (2006). *New Standard Guide for Petrographic Examination of Dimension Stone*. WK2609 (proposed new standard in draft form and under development). Philadelphia, USA.

ASTM International (2008a). *Selection of Dimension Stone for Exterior Use*. ASTM C1528-08. Philadelphia, USA.

ASTM International (2008b). *Standard Specification for Limestone Dimension Stone*. ASTM C568-08a. Philadelphia, USA.

ASTM International (2008c). *Standard Specification for Quartz-Based Dimension Stone*. ASTM C616-08. Philadelphia, USA.

ASTM International (2008d). *Standard Specification for Marble Dimension Stone*. ASTM C503-08a. Philadelphia, USA.

British Standards Institution (1999). *Code of Practice for Site Investigations*: BS 5930. BSI, London.

British Standards Institution (2001a). *Slabs of National Stone for External Paving – Requirements and Test Methods*: BS 1341. BSI, London.

British Standards Institution (2001b). *Setts of Natural Stone for External Paving – Requirements and Test Methods*: BS 1342. BSI, London.

British Standards Institution (2001c). *Kerbs of Natural Stone for External Paving – Requirements and Test Methods*: BS 1343. BSI, London.

British Standards Institution (2002). *Natural Stone. Terminology*: BS EN 12670. BSI, London.

British Standards Institution (2003a). *Natural Stone – Rough Blocks – Requirements*: BS 1467. BSI, London.

British Standards Institution (2003b). *Natural Stone Test Methods – Rough slabs – Requirements*: BS 1468. BSI, London.

British Standards Institution (2004a). *Natural Stone Products – Slabs for Cladding – Requirements*: BS 1469. BSI, London.

British Standards Institution (2004b). *Natural Stone Products – Modular Tiles – Requirements*: BS 12057. BSI, London.

British Standards Institution (2004c). *Natural Stone Products – Slabs for Floors and Stairs –Requirements*: BS 12058. BSI, London.

British Standards Institution (2005). *Specification for Masonry Units – Part 6: Natural Stone Masonry Units*: BS 771-6. BSI, London.

British Standards Institution (2007). *Natural Stone Test Methods – Petrographic Examination*: BS EN 12407. BSI, London.

Burton M (1999). *Designing with Stone.* Ealing Publications Ltd, Berkshire.

Clarke BA (2004). *Clunch the Forgotten Stone: The Conservation of Historic Clunch Buildings in Bedfordshire and Cambridgeshire.* MSc. Thesis, Oxford Brookes University (unpublished).

Dreesen R, Guillite O, Swings J, Nimis PC (1999). Bioreceptivity and biodeterioration of building stones monitored by petrographic techniques. *Proceedings of the 7th Euroseminar on Microscopy Applied to Building Materials.* Delft, Holland, 29 June–2 July 1999, pp. 267–276.

Dunham RJ (1962). Classification of carbonate rocks according to depositional texture. In: Ham, WE (ed). *Classification of Carbonate Rocks.* American Association of Petroleum Geologists, Memoir 1, pp. 108–121.

Folk RL (1959). Practical petrographic classification of limestones. *American Association of Petroleum Geologists, Bulletin* 43:1–38.

Folk RL (1962). Spectral subdivision of limestone types. In: Ham WE (ed). *Classification of Carbonate Rocks.* American Association of Petroleum Geologists, Memoir 1, pp. 62–84.

Howe JA (1910). *The Geology of Building Stones.* Edward Arnold, London.

Ingham JP (2005a). The role of light microscopy in the investigation of historic masonry structures. *Proceedings of the Royal Microscopical Society* 40/1:14–22.

Ingham JP (2005b). Predicting the frost resistance of building stone. *Quarterly Journal of Engineering Geology and Hydrogeology* 38:387–399.

Ingham JP (2007). Assessment of fire-damaged concrete and masonry structures: the application of petrography. *Proceedings of the 11th Euroseminar on Microscopy Applied to Building Materials*, Porto, Portugal, 5–9 June 2007.

International Society for Rock Mechanics (ISRM) (1977). *Rock Characterisation, Testing and Monitoring. Suggested Method for the Petrographic Description of Rock.* Pergamon Press, Oxford.

Jana D (2007). The great pyramid debate: Evidence from detailed petrographic examinations of casing stones from the Great Pyramid of Khufu, a natural limestone from Tura, and a man-made (geopolymeric) limestone. *Proceedings of the 29th Conference on Cement Microscopy*, International Cement Microscopy Association, Quebec City, Canada, pp. 207–266.

Klemm R, Klemm DD (2008). *Stones and Quarries in Ancient Egypt.* The British Museum Press, London.

Mackenzie WS, Donaldson CH, Guildford C (1982). *Atlas of Igneous Rocks and their Textures.* Longman Group UK Limited, Harlow.

Marshall D, Anglin CD, Mumin H (2004). *Ore Mineral Atlas.* Geological Association of Canada – Mineral Deposits Division, St Johns, Newfoundland.

Maxfield V, Peacock D (2001). *The Roman Imperial Quarries – Survey and Excavations at Mons Porphyrities 1994– 1998. Volume 1: Topography and Quarries.* Egypt Exploration Society, London.

Price MT (2007). *Decorative Stone: The Complete Sourcebook.* Thames & Hudson, London.

Smith MR (ed) (1999). *Stone – Building Stone, Rock Fill and Armourstone in Construction.* Engineering Geology Special Publication 16. Geological Society, London.

Studio Marmo (1998). *Natural Stone: A Guide to Selection.* WW Norton & Company, New York.

Terry RD, Chilingar GV (1955). Summary of 'Concerning some additional aids in studying sedimentary formations', by M. S. Shvetsov. *Journal of Sedimentary Petrology* 25:229–234.

Yardley BWD, Mackenzie WS, Guildford C (1990). *Atlas of Metamorphic Rocks and their Textures.* Addison Wesley Longman Limited, Harlow.

## CHAPTER 3

ASTM International (2006). *Standard Specification for Roofing Slate.* ASTM C406-06. Philadelphia, USA.

Bailey D (1994). *The Architect's Guide to Natural Slate Roofing.* Cwt-Y-Bugail Slate Quarries Co Ltd, Blaenau Ffestiniog.

Blanchard I, Sims I (2007). European testing of roofing slate. *Proceedings of the Institution of Civil Engineers: Construction Materials* 160/CM1:1–6.

British Standards Institution (2000). *Slate and Stone Products for Discontinuous Roofing and Cladding – Part 2: Methods of Test*: BS EN 12326-2. BSI, London.

British Standards Institution (2004). *Slate and Stone Products for Discontinuous Roofing and Cladding – Part 1: Product Specification*: BS EN 12326-1. BSI, London.

Coates DT (1993). *Roofs and Roofing: Design and Specification Handbook.* Whittles Publishing, Dunbeath.

Evangelou VP (1995). *Pyrite Oxidation and its Control.* CRC Press, Florida.

Harrison HW (1996). *Roofs and Roofing: Performance, Diagnosis, Maintenance, Repair and the Avoidance of Defects.* BRE Building Elements BR 302. Building Research Establishment, Watford.

Hart D (1991). *The Building Slates of the British Isles.* Report BR195. Building Research Establishment, Watford.

Hunt BJ (1998). Chinese roofing slate – telling the good from the bad. *Roofing Cladding and Insulation*, October, 54–56.

Hunt BJ (2005). Setting new slate standards. *Roofing, Cladding and Insulation*, September, 60–62.

Hunt BJ (2006). Building stones explained 4: Slate. *Geology Today* 22/1:33–40.

Ingham JP (2005). Characterisation of roofing slate by optical microscopy. *Proceedings of the 10th Euroseminar on Microscopy Applied to Building Materials*, Paisley, United Kingdom, 21– 25 June 2005.

Jenkins J (2003). *Slate Roof Bible.* Chelsea Green Publishing Co, White River Jct, Vermont.

Menéndez Seigas JL (1992). *Architecture and Techniques of Roofing Slate*, 2nd edn. Federacian Espanola e la Pizarra, Sobradelo De Valdeorras.

North FJ (1946). *The Slates of Wales*, 3rd edn. National Museum of Wales, Cardiff.

Walsh JA (2002). *Scottish Roofing Slate – Characteristics and Tests.* Historic Scotland Research Report, Edinburgh.

Walsh JA (2003). The relationship between the cleavage properties of natural roofing slate and the thickness of the finished material. *Proceedings of the 9th Euroseminar on Microscopy Applied to Building Materials*, Trondheim, Norway, 9–12 September 2003.

## CHAPTER 4

Abbas A, Fathifazl G, Fournier B, Isgor OB, Zavadil R, Razaqpur AG, Foo S (2007). Quantification of the residual mortar content in recycled concrete aggregates by image analysis. *Proceedings of the 11th Euroseminar on Microscopy Applied to Building Materials*, Porto, Portugal, 5–9 June 2007.

Anon (1995). The description and classification of weathered rock for engineering purposes. Geological Society Engineering Group Working Party Report. *Quarterly Journal of Engineering Geology* 28:207–242.

ASTM International (2008a). *Standard Guide for Petrographic Examination of Aggregates for Concrete*. ASTM C295-08. Philadelphia, USA.

ASTM International (2008b). *Standard Specification for Concrete Aggregates*. ASTM C33/C33M-08. Philadelphia, USA.

British Geological Survey (2002). *Directory of Mines and Quarries*. British Geological Survey, Keyworth.

British Standards Institution (1992). *Specification for aggregates from natural sources for concrete*: BS 882. BSI, London.

British Standards Institution (1994). *Testing Aggregates - Part 104: Method for Qualitative and Quantitative Petrographic Examination of Aggregates*: BS 812. BSI, London.

British Standards Institution (1997). *Tests for General Properties of Aggregates – Part 3: Procedure and Terminology for Simplified Petrographic Description*. BS EN 932-3. BSI, London.

British Standards Institution (1999a). *Code of Practice for Site Investigations*: BS 5930. BSI, London.

British Standards Institution (1999b). *Guide to the Interpretation of Petrographical Examinations for Alkali–Silica Reactivity*: BS 7943. BSI, London.

British Standards Institution (2002a). *Aggregates for Concrete*: BS EN 12620. BSI, London.

British Standards Institution (2002b). *Aggregates for Bituminous Mixtures and Surface Treatments for Roads, Airfields and Other Trafficked Areas*: BS EN 13043. BSI, London.

British Standards Institution (2002c). *Aggregates for Mortar*: BS EN 13139. BSI, London.

British Standards Institution (2002d). *Aggregates for Unbound and Hydraulically Bound Materials for Use in Civil Engineering Work and Road Construction*: BS EN 13242. BSI, London.

British Standards Institution (2002e). *Aggregates for Railway Ballast*: BS EN 13450. BSI, London.

Building Research Establishment (1998). *Recycled Aggregates*. BRE Digest 433. BRE, Watford.

Building Research Establishment (2004). *Alkali-Silica Reaction in Concrete: Detailed Guidance for New Construction*. BRE Digest 330: Part 2. BRE, Watford.

Dolar-Mantuani, L (1983). *Handbook of Concrete Aggregates: A Petrographic and Technological Evaluation*. Noyes Publications, New Jersey.

English Heritage (2000). *The English Heritage Directory of Building Sands and Aggregates*. Donhead Publishing, Shaftesbury.

Fookes PG (1997). Aggregates: a review of prediction and performance. In: *Prediction of Concrete Durability: Proceedings of STATS 21st Anniversary Conference*. E & FN Spon, London, pp. 91–170.

Fookes PG, Lay J, Sims I, West G (eds) (2001). *Aggregates: Sand, Gravel and Crushed Rock Aggregates for Construction Purposes*, 3rd edn. Engineering Geology Special Publication 17. Geological Society, London.

French WJ (1994). Avoiding concrete aggregate problems. In: *Improving Civil Engineering Structures – Old and New*. French WJ (ed). Geotechnical Publishing Ltd, Essex, pp. 65–95.

French WJ (2005). Presidential Address 2003: Why concrete cracks – geological factors in concrete failure. *Proceedings of the Geologists' Association* 116:89–105.

French WJ, Crammond NJ (1980). The influence of serpentinite and other rocks on the stability of concretes in the Middle East. *Quarterly Journal of Engineering Geology* 13:255–280.

Goguel RL, Milestone NB (2007). Alkali release from aggregates: unexplained cracking due to ASR. *Proceedings of the Concrete Platform 2007 Conference*, 19–20 April, pp. 479–487.

Huenger KJ, Weidmueller H (2007). Microscopic investigations on AAR in greywacke aggregates. *Proceedings of the 11th Euroseminar on Microscopy Applied to Building Materials*, Porto, Portugal, 5–9 June 2007.

Hunt BJ (1995). *An Engineering Classification of Quartzite and Chert to Aid Reactivity Assessment*. Paper presented at Concrete Day Southeast, London (not published).

Ingham JP (2005). Investigation of traditional lime mortars – the role of optical microscopy. *Proceedings of the 10th Euroseminar on Microscopy Applied to Building Materials*, Paisley, United Kingdom, 21–25 June 2005.

Jensen V (2007). Reliability of petrographic analysis for aggregates used as building materials. *Proceedings of the 11th Euroseminar on Microscopy Applied to Building Materials*, Porto, Portugal, 5–9 June 2007.

Leemann A, Holzer L. (2001). Influence of mica on the properties of mortar and concrete. *Proceedings of the 8th Euroseminar on Microscopy Applied to Building Materials*, Athens, Greece, 4–7 September 2001, pp. 199–204.

Leslie A, Gibbons P (1999). Technical advice note 19: *Scottish Aggregates for Building Conservation*. Historic Scotland, Edinburgh.

Linqvist JE, Sandström M (1999). Determination of the size distribution, sieve curve, for aggregates using optical microscopy. *Proceedings of the 7th Euroseminar on Microscopy Applied to Building Materials*, Delft, Holland, 29 June–2 July, pp. 297–305.

Midgley HG (1958). The staining of concrete by pyrite. *Magazine of Concrete Research* 10/29:75–78.

Quarry Management (2007/2008). *Directory of Quarries & Quarry Equipment*, 31st edn. QMJ Publishing, Nottingham.

Sherwood PT (2001). *Alternative Materials in Road Construction: a Guide to the Use of Waste, Recycled Materials and By-Products*, 2nd edn. Thomas Telford Publications, London.

Sims I, Smart SAS, Hunt BJ (2002). Practical petrography: the modern assessment of aggregates for alkali-reactivity potential. In: *Industrial Minerals and Extractive Industry Geology*. Scott PW, Bristow CM (eds). The Geological Society, London, pp. 183–188.

Wigum BJ (1996). A classification of Norwegian cataclastic rocks for alkali-reactivity. *Proceedings of the 10th International Conference on Alkali–Aggregate Reaction in Concrete*, Melbourne, Australia, pp. 758–766.

# CHAPTER 5

ASTM International (2004). *Standard Practice for the Petrographic Examination of Hardened Concrete.* ASTM C856-04. Philadelphia, USA.

ASTM International (2007). *Standard Specification for Portland Cement.* ASTM C150-07. Philadelphia, USA.

ASTM International (2008a). *Standard Specification for Concrete Aggregates.* ASTM C33/C33M-08. Philadelphia, USA.

ASTM International (2008b). *Standard Specification for Blended Hydraulic Cement.* ASTM C595-08a. Philadelphia, USA.

ASTM International (2008c). *Standard Test Method for Microscopical Determination of Parameters of the Air-Void System in Hardened Concrete.* ASTM C457-08d. Philadelphia, USA.

ASTM International (2009). *Standard Specification for Ready-Mixed Concrete.* ASTM C94/C94M-09. Philadelphia, USA.

Bamforth PB (2004). *Enhancing Reinforced Concrete Durability: Guidance on Selecting Measures for Minimising the Risk of Corrosion of Reinforcement in Concrete.* Concrete Society Technical Report No. 61. The Concrete Society, Camberley.

Barnes P, Bensted J (eds) (2002). *Structure and Performance of Cements*, 2nd edn. Spon Press, London.

British Standards Institution (1988). *Testing Concrete – Methods for Analysis of Hardened Concrete.* BS 1881: Part 124. BSI, London..

British Standards Institution (1996). *Specification for Sulfate-Resisting Portland Cement.* BS 4027. BSI, London..

British Standards Institution (1999). *Guide to the Interpretation of Petrographical Examinations for Alkali–Silica Reactivity*: BS 7943. BSI, London.

British Standards Institution (2000a). *Cement – Part 1: Composition, Specifications and Conformity Criteria for Common Cements.* BS EN 197-1. BSI, London.

British Standards Institution (2000b). *Concrete – Part 1: Specification, Performance, Production and Conformity.* BS EN 206-1. BSI, London.

British Standards Institution (2002). *Aggregates for Concrete*: BS EN 12620. BSI, London.

British Standards Institution (2005). *Admixtures for Concrete, Mortar and Grout. Part 11: Test Methods. Determination of Air Void Characteristics in Hardened Concrete.* BS EN 480-11. BSI, London.

Bromley A (2002). *A Compendium of Concrete Aggregates used in Southwest England* (unpublished).

Bromley AV, Sibbick RG (1999). The accelerated degradation of concrete in Southwest England. *Proceedings of the 7th Euroseminar on Microscopy Applied to Building Materials*, Delft, Holland, 29 June–2 July, pp. 81–90.

Building Research Establishment (2000). *Corrosion of Steel in Concrete: Investigation and Assessment.* BRE Digest 444: Part 2. BRE, Watford.

Building Research Establishment (2004). *Alkali–Silica Reaction in Concrete: Detailed Guidance for New Construction.* BRE Digest 330: Part 2. BRE, Watford.

Building Research Establishment (2005). *Concrete in Aggressive Ground*, 3rd edn. BRE Special Digest 1, Watford.

Bungey JH, Millard SG, Grantham MG (2006). *Testing of Concrete in Structures*, 4th edn. Routledge, Abingdon.

Bye GC (1999). *Portland Cement.* Thomas Telford Publishing, London.

Campbell DH (1999). *Microscopical Examination and Interpretation of Portland Cement and Clinker*, 2nd edn. Construction Technology Laboratories, a Division of the Portland Cement Association, Skokie, Illinois.

Concrete Society (1987). *Concrete Core Testing for Strength.* Technical Report No. 11. The Concrete Society, Camberley.

Concrete Society (1989). *Analysis of Hardened Concrete – a Guide to Tests, Procedures and Interpretation of Results.* Technical Report No. 32. The Concrete Society, Camberley.

Concrete Society (1992). *Non-Structural Cracking in Concrete.* Technical Report No. 22 3rd edn. The Concrete Society, Camberley.

Concrete Society (1997). *Calcium Aluminate Cements in Construction*: A Reassessment. Technical Report No. 46. The Concrete Society, Camberley.

Concrete Society (1999). *Alkali-Silica Reaction: Minimising the Risk of Damage to Concrete. Guidance Notes and Model Clauses for Specifications.* Technical Report No. 30 3rd edn. The Concrete Society, Camberley.

Concrete Society (2000). *Diagnosis of Deterioration in Concrete Structures.* Concrete Society Technical Report No. 54. The Concrete Society, Camberley.

Concrete Society (2008). *Assessment, Design and Repair of Fire-Damaged Concrete Structures.* Technical Report No. 68. The Concrete Society, Camberley.

Department of the Environment, Transport and the Regions (1999). *The Thaumasite Form of Sulfate Attack: Risks, Diagnosis, Remedial Works and Guidance on New Construction.* Report of the Thaumasite Expert Group, DETR, London.

Dunster A (2002a). *Avoiding Deterioration of Cement Based Building Materials and Components: Lessons from Case Studies 4.* Building Research Establishment, Watford.

Dunster A (2002b). *HAC Concrete in the UK: Assessment, Durability, Management, Maintenance and Refurbishment.* BRE Special Digest 3. Building Research Establishment, Watford.

Eden MA (2003). The laboratory investigation of concrete affected by TSA in the UK. *Cement & Concrete Composites* 25:847–850.

Eden MA, White PS, Wimpenny DE (2007). A laboratory investigation of concrete with suspected delayed ettringite formation – a case study from a bridge in Malaysia. *Proceedings of the 11th Euroseminar on Microscopy Applied to Building Materials.* Porto, Portugal, 5–9 June.

French WJ (1991a). Concrete petrography: a review. *Quarterly Journal of Engineering Geology* 24/1:17–48.

French WJ (1991b). Comments on the determination of the ratio of GGBS to Portland cement in hardened concrete. *Concrete* 25/6:33–36.

French WJ (1992). Determination of the ratio of PFA to Portland cement in hardened concrete. *Concrete* 26/3:43–45.

French WJ (2005). Why concrete cracks – geological factors in concrete failure. *Proceedings of the Geologists' Association* 116:89–106.

Goguel RL, Milestone NB (2007). Alkali release from aggregates: the answer to unexplained cracking due to ASR. *Proceedings of Concrete Platform 2007*, Belfast, pp. 479–487.

Gonçalves D, Rocha F, Salta M, Santos Silva A (2007). Use of microscopical methods to determine the water-cement ratio of hardened concrete. *Proceedings of the 11th Euroseminar on Microscopy Applied to Building Materials.* Porto, Portugal, 5–9 June.

Grove RM (1968). The identification of ordinary Portland cement and sulphate resisting cement in hardened concrete samples. *Silicates Industriels* 10:317–320.

Hewlett PC (ed) (1998). *Lea's Chemistry of Cement and Concrete* 4th edn. Arnold Publishers, London.

Ingham JP (2007). Assessment of fire-damaged concrete and masonry structures: The application of petrography. *Proceedings of the 11th Euroseminar on Microscopy Applied to Building Materials*, Porto, Portugal, 5–9 June 2007.

Ingham JP, Hamm T (2005). Using microscopy and microanalysis to investigate concrete. *Concrete* March:27–29.

Ingham JP, Tarada F (2007). Turning up the heat – full service fire safety engineering for concrete structures. *Concrete* October:27–30.

Institution of Structural Engineers (1992). *Structural Effects of Alkali-Silica Reaction. Technical Guidance on the Appraisal of Existing Structures.* Institution of Structural Engineers, London.

Jakobsen UH, Brown DR, Comeau RJ, Henriksen JHH (2003). Fluorescent epoxy impregnated thin sections prepared for a round robin test on W/C determination. *Proceedings of the 9th Euroseminar on Microscopy Applied to Building Materials*, Trondheim, Norway, 9–12 September.

Jakobsen UH, Pade C, Thaulow N, Brown D, Sahu S, Magnusson O, De Buck S, De Schutter G (2005). The Rapidair system for air void analysis of hardened concrete – a round robin study. *Proceedings of the 10th Euroseminar on Microscopy Applied to Building Materials*, Paisley, United Kingdom, 21–25 June.

Jana D, Erlin B (2007). Carbonation as an indicator of crack age. *Concrete International* May:61–64.

Katayama T, Futagawa T (1997). Petrography of pop-out causing minerals and rock aggregates in concrete – Japanese experience. *Proceedings of the 6th Euroseminar on Microscopy Applied to Building Materials*, Reykjavik, Iceland, 25-27 June, pp. 400–409.

Katayama T (2003). How to identify carbonate rock reactions in concrete. *Proceedings of the 9th Euroseminar on Microscopy Applied to Building Materials*, Trondheim, Norway, 9–12 September.

Katayama T, Tagami M, Sarai Y, Izumi S, Hira T (2003). Alkali-aggregate reaction under the influence of deicing salts in the Hokuriku District, Japan.

*Proceedings of the 9th Euroseminar on Microscopy Applied to Building Materials*, Trondheim, Norway, 9–12 September.

Kay T (1992). *Assessment and Renovation of Concrete Structures.* Longman Scientific & Technical, Harlow.

Lea FM (1970). *The Chemistry of Cement and Concrete*, 3rd edn. Edward Arnold (Publishers) Ltd, London.

Litorowicz A (2005). Influence of crack system quantified by means of optical fluorescent microscopy and image analysis of concrete properties. *Proceedings of the 9th Euroseminar on Microscopy Applied to Building Materials*, Paisley, United Kingdom, 21–25 June.

Metha PK, Monteiro PJM (2006). *Concrete: Microstructure, Properties and Materials*, 3rd edn. McGraw-Hill, New York.

Midgley HG (1958). The staining of concrete by pyrite. *Magazine of Concrete Research* 10/29:75–78.

Neville AM (1995). *Properties of concrete*, 4th edn. Longman Group Limited, Harlow.

Neville AM (2003). How closely can we determine the water-cement ratio of hardened concrete? *Materials and Structures* 36:311–318.

NT Build 361 (1999). *Concrete, Hardened: Water-Cement Ratio.* Nordtest Method, 2nd edn.

Palmer D (1992). *The Diagnosis of Alkali-Silica Reaction.* Report of a working party. British Cement Association, Wexham Springs, Slough.

Plum DR, Hammersley GP (1984). Concrete attack in an industrial environment. *Concrete* 18/5:8–11.

Pomeroy RD (1992). *The Problem of Hydrogen Sulphide in Sewers*, 2nd edn. Clay Pipe Development Association Limited, London.

Quillin K (2001). Delayed ettringite formation: *in situ* concrete. *BRE Information Paper IP11/01*. Building Research Establishment, Watford.

Ravenscroft PE (1982). Determining the degree of hydration of hardened concrete. *Forum* Jan/Feb:4.

Rushton T (2006). *Investigating Hazardous & Deleterious Building Materials.* RICS Books, London.

Sibbick RG, Crammond NJ (1999). Two case studies into the development of the thaumasite form of sulfate attack (TSA) in hardened concretes. *Proceedings of the 7th Euroseminar on Microscopy Applied to Building Materials*, Delft,

Holland, 29 June-2 July, pp. 203–212.

Sibbick T, Crammond NJ (2003). The petrographic examination of popcorn calcite deposition (PCD) within concrete mortar, and its association with other forms of degradation. *Proceedings of the 9th Euroseminar on Microscopy Applied to Building Materials*, Trondheim, Norway, 9–12 September.

Sibbick T, Brown D, Dragovic B, Knight C, Garrity S, Comeau R (2007). Investigation into the determination of water to cementitious (W-CM) binder ratios by the use of fluorescent microscopy technique in hardened concrete samples: Part 1. *Proceedings of the 11th Euroseminar on Microscopy Applied to Building Materials*, Porto, Portugal, 5–9 June.

Sims I, Hunt BJ, Miglio BF (1992). Quantifying microscopical examinations of concrete for alkali aggregate reactions (AAR) and other durability aspects. In: *Durability of Concrete*, Holm J (ed). SP131, American Concrete Institute, Farmington Hills, pp. 267–287.

Sims I, Ingham JP, Sotiropoulos P (2004). Forensic petrography and AAR diagnosis. *Proceedings of the 12th International Conference on Alkali-Aggregate Reactions*, Beijing, China, 15-19 October, pp. 995–1004.

Skalny J, Marchand J, Odler I (2002). *Sulfate Attack on Concrete.* Spon Press, London.

Smart S (1999). Concrete – the burning issue. *Concrete* July/August:30–34.

St John DA, Poole AB, Sims I (1998). *Concrete Petrography, a Handbook of Investigative Techniques.* Edward Arnold (Publishers) Ltd, London.

Stanton T (1940). Expansion of concrete through reaction between cement and aggregate. *Proceedings American Society of Civil Engineers* 66:1781–1811.

Stimson CC (Chairman) (1997). *The 'Mundic' Problem – a Guidance Note*, 2nd edn. Royal Institution of Chartered Surveyors, London.

Taylor HFW (1997). *Cement Chemistry.* Thomas Telford, London.

Thomas M, Folliard K, Drimalas T, Ramlochan T (2007). Diagnosing delayed ettringite formation in concrete structures. *Proceedings of the 11th Euroseminar on Microscopy Applied to Building Materials*. Porto, Portugal, 5–9 June.

Walker HN, Lane DS, Stutzman PE (2006). *Petrographic Methods of Examining Hardened Concrete: A Petrographic Manual* (Revised 2004). National

Technical Information Service, Springfield.

West G (1996). *Alkali-Aggregate Reaction in Concrete Roads and Bridges*. Thomas Telford, London.

Wong HS, Buenfeld NR (2007). Estimating the water/cement (W/C) ratio from the phase composition of hardened cement paste. *Proceedings of the 11th Euroseminar on Microscopy Applied to Building Materials*, Porto, Portugal, 5–9 June.

Yan L, Lee CF, Pei-xing F (2004). Alkali-silica reaction (ASR) characteristics of concrete made from granite aggregates. *Proceedings of the 12th International Conference on Alkali-Aggregate Reaction in Concrete*, Beijing, China, pp. 758–766.

## CHAPTER 6

ASTM International (2004). *Standard Practice for the Petrographic Examination of Hardened Concrete*. ASTM C856-04. Philadelphia, USA.

ASTM International (2005). *Standard Specification for Calcium Silicate Brick (Sand-Lime Brick)*. ASTM C73-05. Philadelphia, USA.

Aircrete Products Association (2005). *Code of Best Practice for the Use of Aircrete Products*. AACPA, United Kingdom.

Bowley, B (1993/4). Calcium silicate bricks. *Structural Survey* 12/6:16–18.

British Standards Institution (2003). *Specification for masonry units. Calcium silicate masonry units*. BS EN 771-2. BSI, London.

British Standards Institution (2005). *Fly Ash for Concrete – Part 1: Definitions, Specifications and Conformity Criteria*. BS EN 450-1. BSI, London.

British Standards Institution (2008). *Specification for Cast Stone*. BS 1217. BSI, London.

Building Research Establishment (1992). *Calcium Silicate (Sandlime, Flintlime) Brickwork*. BRE Digest 157. BRE, Watford.

Building Research Establishment (2002a). *AAC 'Aircrete' Blocks and Masonry* (in two parts). BRE Digest 468. BRE, Watford.

Building Research Establishment (2002b). *Reinforced Autoclaved Aerated Concrete Panels: Review of Behaviour, and Developments in Assessment and Design*. BR 445. BRE, Watford.

Cast Stone Institute (2005). *Technical Manual with Case Histories*, 4th edn. Lebanon, Pennsylvania, USA.

*Control of Asbestos at Work Regulations*

*2002*. Statutory Instrument 2002 No. 2675. The Stationary Office, United Kingdom.

Dawson S (2003). *Cast in Concrete – a Guide to the Design of Precast Concrete and Reconstructed Stone*. Architectural Cladding Association, Leicester, United Kingdom.

Health and Safety Executive (2005). *Asbestos: The Analysts' Guide for Sampling, Analysis and Clearance Procedures*. HSE Books, United Kingdom.

Levitt M (1982). *Precast Concrete: Materials, Manufacture, Properties and Usage*. Applied Science Publishers, London.

Richardson JG (1991). *Quality in Precast Concrete: Design, Productions, Supervision*. Longman Scientific & Technical, Harlow.

United Kingdom Cast Stone Association (2004). *Technical Manual for Cast Stone*. UKCSA, Crowthorne.

## CHAPTER 7

ASTM International (2004). *Standard Practice for the Petrographic Examination of Hardened Concrete*. ASTM C856-04. Philadelphia, USA.

British Standards Institution. (1989). *Workmanship on Building Sites: Code of Practice for Wall and Floor Tiling – Ceramic Tiles, Terrazzo Tiles and Mosaics*: BS 8000-11.1. BSI, London.

British Standards Institution (2002). *Screed Material and Floor Screeds – Screed Material – Properties and Requirements*: BS EN 13813. BSI, London.

British Standards Institution (2003). *Screed Bases and In-Situ Floorings – Part 1: Concrete Bases and Cement Sand Levelling Screeds to Receive Floorings – Code of Practice*: BS 8204-1. BSI, London.

British Standards Institution (2004a). *Terrazzo Tiles: Terrazzo Tiles for Internal Use*: BS EN 13748-1. BSI, London.

British Standards Institution (2004b). *Terrazzo Tiles: Terrazzo Tiles for External Use*: BS EN 13748-2. BSI, London.

British Standards Institution (2009). *Wall and Floor Tiling. Design and Installation of Terazzo, Natural Stone and Agglomerated Stone Tile and Slab Flooring. Code of Practice*: BS 8204-1. BSI, London.

Building Research Station (1961). *Principals of Modern Building, Volume 2: Floors and Roofs*. HMSO, London.

Carpenter J, Lazarus D, Perkins C (2006). *Safer Surfaces to Walk on: Reducing the Risk of Slipping*. Report C652. CIRIA, London.

Chaplin RG (1997). *Floor Levelling Screeds*. BCA Guide. British Cement Association, Crowthorne.

EFNARC (2001). *Specifications and Guidelines for Synthetic Resin Flooring*. EFNARC, Aldershot.

Fawcett J (1998). *Historic Floors: Their History and Conservation*. Butterworth-Heinemann Ltd., Oxford.

Gatfield MJ (1998). *Screeds, Flooring and Finishes: Selection, Construction and Maintenance*. Report R184. CIRIA, London.

Perkins PH (1993). *Concrete Floors, Finishes and External Paving*. Butterworth-Heinemann, Oxford.

Pye PW, Harrison HW (1997). *Floors and Flooring: Performance, Diagnosis, Maintenance, Repair and the Avoidance of Defects*. BRE Building Elements BR 332. BRE, Watford.

Ripley J (2005). *Ceramic and Stone Tiling: A Complete Guide*. The Crowood Press Ltd, Ramsbury.

Stone Federation of Great Britain (2000). *Natural Stone Flooring: Code of Practice for the Design and Installation of Internal Flooring*. SFGB, London.

Wright A (1999). Care and Repair of Old Floors. Technical Pamphlet 15. Society for the Protection of Ancient Buildings, London.

## CHAPTER 8

Adam JP (1999). *Roman Building: Materials and Techniques*. Routledge, London.

Allen G, Allen J, Elton N, Farey M, Holmes S, Livesey P, Radonjic M (2003). *Hydraulic Lime Mortar for Stone, Brick and Block Masonry*. Donhead Publishing Ltd, Shaftesbury.

Arnold D (1991). *Building in Egypt: Pharaonic Stone Masonry*. Oxford University Press, Oxford.

Ashurst J (2002). *Mortars, Plasters and Renders in Conservation*, 2nd edn. Ecclesiastical Architects and Surveyors Association, London.

ASTM International (2003). *Standard Specification for Finishing Hydrated Lime*. ASTM C206-03. Philadelphia, USA.

ASTM International (2004). *Standard Practice for the Petrographic Examination of Hardened Concrete*. ASTM C856-04. Philadelphia, USA.

ASTM International (2005a). *Standard Specification for Gypsum Plasters.* ASTM C28/C28M-00 (2005). Philadelphia, USA.

ASTM International (2005b). *Standard Test Method for Examination and Analysis of Hardened Masonry Mortar.* ASTM C1324-05. Philadelphia, USA.

ASTM International (2006a). *Standard Specification for Hydrated Lime for Masonry Purposes.* ASTM C207-06. Philadelphia, USA.

ASTM International (2006b). *Standard Specification for Application of Portland Cement-Based Plaster.* ASTM C926-06. Philadelphia, USA.

ASTM International (2008a). *Standard Specification for Mortar for Unit Masonry.* ASTM C270-08a. Philadelphia, USA.

ASTM International (2008b). *Standard specification for lime putty for structural purposes.* ASTM C1489-01(2008)e1. Philadelphia, USA.

Barnes P, Bensted J (eds) (2002). *Structure and Performance of Cements*, 2nd edn. Spon Press, London.

Beare M (2004). Building commercially with lime mortar: Justifying what we know will work – the engineering aspects. *Journal of the Building Limes Forum* 11:15–22.

Blezard RG (1998). The history of calcareous cements. In: *Lea's Chemistry of Cement and Concrete*, 4th edn. Hewlett PC (ed). Arnold, London, pp. 1–23.

British Gypsum (2005). *The White Book.* British Gypsum, East Leake, UK.

British Standards Institution (1991). *Code of Practice for External Renderings*: BS 5262. BSI, London.

British Standards Institution (1995). *Code of Practice for Dry Lining and Partitioning using Gypsum Plasterboard*: BS 8212. BSI, London.

British Standards Institution (1997). *Products and Systems for the Protection and Repair of Concrete Structures: Definitions, Requirements, Quality Control and Evaluation Of conformity – General Principles for the Use of Products and Systems*: DD ENV 1504-9. BSI, London.

British Standards Institution (1998). *The Principles of the Conservation of Historic Buildings.* BS 7913. BSI, London.

British Standards Institution (2000). *Code of Practice for Cleaning and Surface Repair of Buildings – Part 2: Surface Repair of Natural Stones, Brick and Terracotta.* BS 8221. BSI, London.

British Standards Institution (2001). *Building Lime – Part 1: Definitions, Specifications and Conformity Criteria.* BS EN 459-1. BSI, London.

British Standards Institution (2002). *Specification for Mortar for Masonry – Part 2: Masonry Mortar*: BS EN 998-2. BSI, London.

British Standards Institution (2003). *Specification for Mortar for Masonry – Part 1: Rendering and Plastering Mortar*: BS EN 998-1. BSI, London.

British Standards Institution (2004a). *Masonry Cement – Part 1: Composition, Specifications and Conformity Criteria*: BS EN 413-1. BSI, London.

British Standards Institution (2004b). *Gypsum Plasterboards – Definitions, Requirements and Test Methods*: BS EN 520. BSI, London.

British Standards Institution (2005a). *Mortar – Methods of Test for Mortar. Chemical Analysis and Physical Testing*: BS 4551. BSI, London.

British Standards Institution (2005b). *Design, Preparation and Application of External Rendering and Internal Plastering. External Rendering*: BS EN 13914-1. BSI, London.

British Standards Institution (2005c). *Gypsum Binders and Gypsum Plasters – Definitions and Requirements*: BS EN 13279-1. BSI, London.

Building Research Establishment (1976). *External Rendered Finishes.* BRE Digest 196. BRE Watford.

Building Research Establishment (1986). *Site-Applied Adhesives – Failures and How to Avoid Them. BRE Information Paper 12/86.* BRE, Watford.

Building Research Establishment. (1991). *Building mortar.* BRE Digest 362. BRE, Watford.

Carò F, Di Giulio A, Marmo R (2006). Textural analysis of ancient plasters and mortars: reliability of image analysis approaches. In: *Geomaterials in Cultural Heritage.* Maggetti M, Messiga B (eds). Special Publication 257, Geological Society, London, pp. 337–345.

Concrete Repair Association (2001). *The Route to a Successful Concrete Repair.* CRA, Farnham.

Concrete Society (1984). *Repair of Concrete Damaged by Reinforcement Corrosion.* Concrete Society Technical Report No. 26. The Concrete Society, London.

Concrete Society (1994). *Polymers in Concrete.* Concrete Society Technical Report No. 39. The Concrete Society, London.

Concrete Society (1997). *Calcium Aluminate Cements in Construction: A Re-assessment.* Concrete Society Technical Report No. 46. The Concrete Society, London.

Concrete Society (2005). *Mortars for Masonry: Guidance of Specifications, Types, Production and Use.* Concrete Society Good Concrete Guide No. 4. Concrete Society, London.

Cowper AD (1927). *Lime and Lime Mortars.* Building Research Station, Garston.

Davey N (1961). *A History of Building Materials.* Phoenix House, London.

English Heritage (2001). *Anthrax and Historic Plaster: Managing Minor Risks in Historic Building Refurbishment.* Technical Advice Note, English Heritage, London.

Francis AJ (1977). *The Cement Industry 1796–1914: A History.* David & Charles, Newton Abbot.

Green GW (1984). Gypsum analysis with the polarising microscope. In: *The Chemistry and Technology of Gypsum.* Kuntze RA (ed). ASTM STP 861. ASTM International, Philadelphia, pp. 22–47.

Hansen EF (2005). Technology used in the production of ancient Maya mortars and plasters. *Proceedings of the International Building Lime Symposium 2005.* Orlando, Florida.

Holmes S, Wingate M (2002). *Building with Lime: A Practical Introduction*, revised edition. ITDG Publishing, London.

Hughes JJ, Leslie AB, Callebaut K (2001). The petrography of lime inclusions in historic lime based mortars. *Proceedings of the 8th Euroseminar on Microscopy Applied to Building Materials.* Athens, Greece, 4–7 September 2001, pp. 359–364.

Hughes JJ, Válek J (2003). *Mortars in Historic Buildings: A Review of the Conservation, Technical and Scientific Literature.* Historic Scotland, Edinburgh.

Hunt BJ, Everitt F (2001). Analysing modern concretes. *Concrete* 35/1:40–42.

Induni B (2005). The strange history of the lime revival. *Proceedings of the International Building Lime Symposium 2005.* Orlando, Florida.

Ingham JP (2003). Laboratory investigation of lime mortars, plasters and renders. *Journal of the Building Limes Forum* 10:17–36.

Ingham JP (2004). *Investigation of Roman Lime-Based Construction Materials from Archaeological Investigations at Verulamium.* Report prepared for the St Albans & Hertfordshire Architectural and Archaeological Society (not published).

Ingham JP (2005). Investigation of traditional lime mortars: the role of optical microscopy. *Proceedings of the 10th Euroseminar on Microscopy Applied to Building Materials*, Paisley, United Kingdom, 21–25 June 2005.

Jensen AD (1997). Microscopy of cementitious repair mortars. *Proceedings of the 6th Euroseminar on Microscopy Applied to Building Materials*, Reykjavik, Iceland, 25–27 June 1997, pp. 217–219.

Kelsall F (1989). Stucco. In: *Good and Proper Materials – the Fabric of London since the Great Fire*. Hobhouse H, Saunders A (eds). The London Topographical Society Publication No.140, London, pp. 18–24.

Kighelman J, Scrivener K, Zurbriggen R, de Gasparo A, Herwegh M (2005). Influence of additives on phase composition and microstructures of mixed-binder based self-levelling floor compounds (SLC). *Proceedings of the 10th Euroseminar on Microscopy Applied to Building Materials*, Paisley, United Kingdom, 21–25 June 2005.

Kuntze RA (ed) (1984). *The Chemistry and Technology of Gypsum*. ASTM STP 861 ASTM, Philadelphia.

Leslie AB, Hughes JJ (2002). Binder microstructure in lime mortars: implications for the interpretation of analysis results. *Quarterly Journal of Engineering Geology and Hydrogeology* 35:257–263.

Lindqvist JE, Johansson S (2007). Aggregate shape and orientation in historic mortars. *Proceedings of the 11th Euroseminar on Microscopy Applied to Building Materials*, Porto, Portugal, 5–9 June 2007.

Mertens G, Elson J (2005). Evaluation of the determination of the grain size distribution of sands used in mortars by computer assisted image analysis. *Proceedings of the 10th Euroseminar Applied to Building Materials*, Paisley, United Kingdom, 21–25 June 2007.

Middendorf B (2001). Crystallization and recrystallization process of gypsum mortars during weathering. *Proceedings of the 8th Euroseminar on Microscopy Applied to Building Materials*, Athens, Greece, 4–7 September 2001, pp. 365–372.

Millar W (1998). *Plastering Plain and Decorative* (reprinted edition). Donhead Publishing, Shaftesbury. Originally published in 1897.

Monks B (2000). *Rendering – a Practical Handbook*. Concrete Society Good Concrete Guide No. 3. Concrete Society, London.

Munsell™ Soil Color Charts (1994). *Revised Edition Soil Color Charts*. Macbeth Division of Kallmorgan Instruments Corporation, New Windsor.

Odler I (2000). *Special Inorganic Cements*. Spon Press, London.

Oleson JP, Brandon C, Cramer SM, Cucitore R, Gotti E, Hohlfelder RL (2004). The ROMACONS project: a contribution to the historical and engineering analysis of hydraulic concrete in Roman maritime structures. *The International Journal of Nautical Archaeology*, 33/2:199–229.

Pettijohn FJ, Potter PE, Siever R (1987). *Sand and Sandstone*. Springer-Verlag, New York.

Prakaypun W, Jinawath S (2003). Comparative effect of additives on the mechanical properties of plasters made from flue-gas desulfurized and natural gypsums. *Materials and Structures* 36/255:51–58.

Pritchett I (2005). Lime and low-energy masonry. *Journal of the Building Limes Forum* 12:50–61.

Radonjic M, Allen G, Livesey P, Elton N, Farey M, Holmes S, Allen J (2001a). ESEN characterisation of ancient lime mortars. *Journal of the Building Limes Forum* 8:38–49.

Radonjic M, Hallam KR, Allen GC, Hayward RJ (2001b). The mechanism of carbonation in lime-based mortars. *Journal of the Building Limes Forum* 8:50–64.

RILEM Draft Recommendation (2001). Rilem TC 167-Con: Characterisation of old mortars. *Materials and Structures* 34:387–388.

Siddall R (2000). The use of volcaniclastic material in Roman hydraulic concretes: a brief review. In: *The Archaeology of Geological Catastrophes*. McGuire WG *et al.* (eds). Special Publications, 171, Geological Society, London, pp. 339–344.

Stefanidou M (2000). Estimation of porosity by using different techniques. *Proceedings of the 8th Euroseminar on Microscopy Applied to Building Materials*, Athens, Greece, 4–7 September 2001, pp. 619–624.

Swallow P, Carrington D (1995). Lime and lime mortars – part one. *Journal of Architectural Conservation* 1/3:7–25.

Verrall W (1997). *The Modern Plasterer* (reprinted edition). Donhead Publishing, Shaftesbury. Originally published in the early 1930s.

Vicat LJ (1837). *A Practical and Scientific Treatise on Calcareous Mortars and Cements, Artificial and Natural*. John Weale, London.

Vitruvius P (1960). *The Ten Books on Architecture*. Translated by Morgan HM. Dover Publications, New York.

Watson J (1922). *Cements and Artificial Stone: A Descriptive Catalogue of the Specimens in the Sedgwick Museum, Cambridge*. W. Heffer & Sons Ltd, Cambridge.

# CHAPTER 9

Ashurst J, Ashurst N (1988). *Brick, Terracotta and Earth*. Practical Building Conservation: English Heritage Technical Handbook, Volume 2. Gower Technical Press, Aldershot.

British Standards Institution (1995). *Glass for Glazing – Part 1: Classification*: BS 952-1. BSI, London.

British Standards Institution (2003). *Specification for Masonry Units. Clay Masonry Units*: BS EN 771-1. BSI, London.

British Standards Institution (2007). *Natural Stone Test Methods – Petrographic Examination*: BS EN 12407. BSI, London.

Campbell WP, Pryce W (2003). *Brick: A World History*. Thames & Hudson Ltd, London.

Chinn RE (2003). *Ceramography: Preparation and Analysis of Ceramic Microstructures*. ASM International, Ohio.

Dunham AC (1992). Developments in industrial mineralogy: I. The mineralogy of brick-making. *Proceedings of the Yorkshire Geological Society* 49/2:95–104.

Hernández-Crespo MS, Romero M, Rincón JM (2007). Microstructure and microanalytical characterization of ceramic bricks. *Proceedings of the 11th Euroseminar on Microscopy Applied to Building Materials*, Porto, Portugal, 5–9 June 2007.

Insley H, Fréchette van D (1955). *Microscopy of Ceramics and Cements Including Glasses, Slags and Foundry Sands*. Academic Press, New York.

Jornet A, Romer B (1999). Loss of adhesion of ceramic tiles in a floor: A case study. *Proceedings of the 7th Euroseminar on Microscopy Applied to Building Materials.* Delft, Holland, 29 June–2 July, pp. 399–405.

Kopar T, Ducman V (2005). Low-vacuum SEM analysis of ceramic tiles with emphasis on glaze defects characterisation. *Proceedings of the 10th Euroseminar on Microscopy Applied to Building Materials.* Paisley, United Kingdom, 21–25 June.

Lynch G (1994). *Brickwork: History, Technology and Practice. Volumes 1 and 2.* Donhead Publishing Ltd, London.

Middleton A, Freestone I (eds) (1991) *Recent Developments in Ceramic Petrology.* British Museum Occasional Paper Number 81. British Museum Press, London.

Papayianni I, Stefanidou M (2003). Xenoliths in the bricks of ancient technology. *Proceedings of the 9th Euroseminar on Microscopy Applied to Building Materials.* Trondheim, Norway, 9–12 September.

Pavía S (2006). The determination of brick provenance and technology using analytical techniques from the physical sciences. *Archaeometry* 48/2:201–218.

Prentice JE (1990). Structural clay products. In: *Geology of Construction Materials.* Chapman and Hall, London, pp. 139–170.

Reedy CL (2008). *Thin-Section Petrography of Stone and Ceramic Cultural Materials.* Archetype Publications Ltd, London.

Reeves GM, Sims I, Cripps JC (eds) (2006). *Clay Materials Used in Construction.* Engineering Geology Special Publication 21. Geological Society, London.

Rigby GR (1948). *The Thin-Section Mineralogy of Ceramic Materials.* The British Refractories Research Association, Stoke-on-Trent.

Stokes NC (1998) *The Glass and Glazing Handbook.* Standards Australia; SAA HB125-1998.

Stratton M (1994). *The Terracotta Revival: Building Innovation and the Image of the Industrial City in Britain and North America.* Weidenfeld & Nicolson, London.

Teutonico JM (ed) (1996). *Architectural Ceramics: Their History, Manufacture and Conservation.* A joint symposium of English Heritage and the United Kingdom Institute for Conservation. Biddles Ltd, UK.

Warren J (1999). *Conservation of Brick.* Butterworth-Heinemann, Oxford.

Wong CH, Chen HF, Tay YPK (2003). Diagnosing tiling failures with the aid of microscopic and macroscopic examination. *Proceedings of the 9th Euroseminar Applied to Building Materials*, Trondheim, Norway, 9–12 September.

# CHAPTER 10

ASTM International (2001). *Standard Specification for Hot-Mixed, Hot-Laid Bituminous Paving Mixtures.* ASTM D3515-01. Philadelphia, USA.

ASTM International (2004). *Standard Practice for the Petrographic Examination of Hardened Concrete.* ASTM C856-04. Philadelphia, USA.

British Aggregate Construction Materials Industries (1992). *Bituminous Mixes and Flexible Pavements: An Introduction.* BACMI, London.

British Standards Institution (2005a). *Hot Rolled Asphalt for Roads and Other Paved Areas. Part 1 - Specification for Constituent Materials and Asphalt Mixtures*: BS 594-1. BSI, London.

British Standards Institution (2005b). *Coated Macadam (Asphalt Concrete) for Roads and Other Paved Areas. Part 1 - Specification for Constituent Materials and for Mixtures*: BS 4987-1. BSI, London.

British Standards Institution (2007). *Natural Stone Test Methods – Petrographic Examination*: BS EN 12407. BSI, London.

Broekmans MATM (2005). Failure of greenstone, jasper and mylonite aggregate in asphalt due to studded tyres: similarities and differences. *Proceedings of the 10th Euroseminar on Microscopy Applied to Building Materials.* Paisley, United Kingdom, 21–25 June.

Eriksen K (1999). Microscopy in roads. *Proceedings of the 7th Euroseminar on Microscopy Applied to Building Materials.* Delft, Holland, 29 June–2 July, pp. 377–380.

Fookes PG, French WJ (1977). Soluble salt damage to surfaced roads in the Middle East. *The Highway Engineer* December:10–20.

Hunter RN (ed) (1994). *Bituminous Mixtures in Road Construction.* Thomas Telford, London.

Schlangen E (1999). Characterisation of aggregate distribution, size and shape and determination of mechanical properties of asphalt by using image techniques. *Proceedings of the 7th Euroseminar on Microscopy Applied to Building Materials.* Delft, Holland, 29 June–2 July, pp. 381–386.

West G (1995). Petrographical examination of road construction materials. In: *Engineering Geology of Construction.* Eddleston M, Walthall S, Crips JC, Culshaw MG (eds). Engineering Geology Special Publication No. 10, Geological Society, London, pp. 245–251.

Whiteoak D, Read J (2003). *The Shell Bitumen Handbook.* Thomas Telford Ltd, London.

# Index

Note: Page numbers are shown in light type, with those in *italic* referring to tables. Figure numbers are shown in **bold** type

AAR, *see* alkali-aggregate reaction (AAR)
Aberllefenni slate 103
Acetic acid 108, 214
Acid attack 107–109, 214–215
ACR, *see* alkali-carbonate reaction
Additions, mineral 86–88
Additives, plasters 142
Adhesive 160–162, 168–170, 330–336, 359
Admixtures 86, 89, 92, 151, 152, 155, 157, 159
Aerated concrete 124, 125
Aggregates 7
   alkali-reactivity for concrete 69–73
   artificial 61, 62, 78, 127, 140–141, 271–273
   fines 68, 139, 154, 311
   grading 64–66, 78–79, 130–133
   lime mortars 144, 279–284
   low-density 62, 127, 141–142, 274–275
   matching for conservation 74
   petrographic examination 16, *16*, 61
   Portland cement mortars 154, 279, 309–311
   recycled 64
   road 61, 66, 172–174, 363–369
   soundness and impurities 66–69, 135–141
   types and properties 61–63, 76–78
   UK production *8*
Air content 83, 186–187
Air entrainment 89, 92, 124, 185, 320–321
Air pollution 47
Air voids 92–93, 184–185, 187, 188
   Portland cement mortar 156–158, 320–323
Aircrete products 121, 124, 243–245

Aksum 27, 28
Alabaster 43, **78**
Alkali attack 109
Alkali-aggregate reaction (AAR) 69–73, 109–114, *111*, 218–226
Alkali-carbonate reaction (ACR) 113
Alkali–silica gel 100, 178, 219–225
Alkali-silica reaction (ASR) 70, *70*, 110–113, 218–226
Alkaline hydrolysis 109, 119–120, **237**
Alkalis 69, 109, 135
Allochems 34, 46, 65
Alta Quartzite 45, 81
Aluminate phases 79, *81*, 84, 169
Amphibolite *44*, 65, *70*
Anatase 52
Ancaster limestone 36, 49
Andesite *22*, *25*, 136
Anhydrite 139, 269, 334
Anhydrite screed 130, 257–259
Animal hair 42, 148, 297
Anstrude limestone 39, 61
Apatite 52
Aragonite 212
Architectural glass *16*, 163, 170, 360–362
Argentina 107
Armourstone 7
Artificial aggregates 61, 62, 78, 127, 140–141, 271–273
Asbestos 31, 121
Asbestos cement products 126–127, 252–253
Asbestos fibres 121, 127, 252–253
Asphalt *129*, 171, 172, 174, 366
Aspin, John 75
ASR, *see* alkali-silica reaction
ASTM International 9, 22–23
Aswan granite 20, 26
Augite 29, 36
Autoclaved aerated concrete (AAC/aircrete) 124, 243–245
Azul de Macaubas 45, 82

Bacoli pozzolana 86, 173
Bacteria 56, 108
Baltic Brown granite 23
Barrington clunch 39, 60
Basalt *22*, *25*, 30, 36

Bassanite 139, 140, 267–268, 290
Bath limestone 37, 52
Beach sand 62, 122, 144, 284
Belgium 40, 63
Belite 79, 146, 169, 285
Bioclasts 34, 56–57, 61–63
Biodeterioration 48, 93
Biotite mica 20–22, 26, 31
Birefringence imaging microscope 11–12, 14
Bituminous mixtures *16*, 171–174, 364–370
Bituminous sandstone 68, 140
'Blacktop' 171
Blaenau Ffestiniog slate 102
Blastfurnace cement 88
Blastfurnace slag *87*, 175
Bleeding 90, 94, 190, 197
Blended cements 79, 86, 88–89
Blue Pearl syenite 26, 28
Blue Pennant sandstone 33, 43
Boiler clinker 128, 161, 304
Brazil 45, 57–58
Brick
   calcium silicate 124, 126, 246–251
   deterioration 166, 349–350
   fired clay 7, 164–166, 338–348
Brick dust 148, 294
Brickwork mortar *137*, 144, 156, 279, 318, 325
British Standards Institution 9, 24
Brownmillerite 83
Brucite 106, 107, 113, 212
Building stone
   classification 21, *22*
   decay 47–50, 90–97
   defects 47, 88–89
   heating 48–50, *50*, 95–97
   history of use 21
   igneous rocks 20–40, 24–31, *25*
   metamorphic rocks *22*, 44–46, 79–87
   petrographic examination 24
   sampling requirements *16*
   sedimentary rocks *22*, 32–43, 41–78
   testing 22–23, *23*, 24
   UK production *8*
Bulgaria 89

Buxy limestone 39, 62

Caithness flagstone 34, 45
Calcination 50, 62, 116, 152
Calcined flint aggregate 129, 160, 249, 341
Calcite 101, 201, 303
   staining techniques 177
Calcite veins 55, 106, 109, 110, 115
Calcium aluminate cement 117, 159
Calcium aluminoferrites 84–85, 167
Calcium bicarbonate 151
Calcium carbonate 124, 143
Calcium hydroxide 86, 172
Calcium oxide (CaO/quicklime) 86, 124, 143
Calcium silicate bricks 115, 124–126, 246–251
Calcium silicate hydrate gel 86, 174
Calcium silicate hydrates 85–86, 124
Calcium sulfate screed 130, 257–259
Caliza Capri limestone 40, 65
Cambridgeshire Clunch 39, 60
Cambusmore sandstone 33, 44
Carbonation 17, 96–97, 124, 151, 193–194
Carboniferous limestone 35, 40, 46, 63
Caribbean 62, 66
Carrara marble 45–46, 83–84
Cast stone 121, 122–124, 238–242
Castilla slate 108
Cataclasite 148–149
Cement 75
   chemists' notation *82*
   types 79–86, *80–81*
   UK production *8*
Cement balls 122, 130, 156, 238, 255, 319
Cement grains 82, 86, 162–171
Cement lime mixtures 138
Cement paste 85, 172
Cement replacement materials 86, 158–159
Cenospheres 87, 88, 244–245
Ceramic tiles 7, *16*, 163, 168–170, 357–359

Ceramics 163
Chalcedony 70, *70*, 143
Chalk 38–39, 58, 138, 281
Charcoal 148, 295
Charnokite 45, 80
Chemical additives, plaster 142
Chemical admixtures 86, 89, 92, 151, 152, 155, 157, 159
Chemical attack 107–109
Chert 43, 72, *72*, **77**, 145
China 21–22, 33, 35, 86, 116–117
    Great Wall 148–150, 164, 298, 338–340
China clay waste 105, 208
Chloride ingress 85
Chlorides 68, 98
Chlorite 30, *50*, 52, 98, 100–101
Chloritoid 52
Chrysotile 40, 127, 252–253
Cladding 7, 21, 23, 29, 35, 48–49, 62, 94
Clay 137, 164
Clay brick 164–166, 338–343
Clay minerals 43, 88, 164
    staining 178
Cleavage, slate 51, 53–54, 102–105
Clinker
    cement 79, *81*, 162–169
    furnace 16, 78, *80*, 161, 304
Clinopyroxene 31–32
Clipsham limestone 51
Clunch limestone 38–39, 59–60
Collophane 59–60
Colorants, *see* Pigments
Compaction 92–93, 150, 300
Concrete 7, 75
    aggregate types 76–78
    assessment of structures 75–76
    deterioration/damage 96–102
    fire damage 114–117, *114*, *116*, 227–231
    reinforcement 75, 97–98
    UK production *8*
    water:cement ratio 89–92, 179–182
    workmanship 94–95, 189–192
Concrete products 121–122
    aircrete 124, 243–245
    architectural cast stone 122–124, 238–242
    asbestos cement 126–127
    calcium silicate units 124–126
    petrographic examination 122
Concrete Society, The 75, 92, 94,

98, 110, *114*
Conservation 74, 138
Conversion 117
Coral aggregate 62, 126
Core samples 13, 16–17, 189
Cornwall 39, 58
Corrosion, steel reinforcement 97–98, 195
Cortex, flint 67, 137
Cracks and cracking 98–100, *99*, 196–198
    dating 97
Crazing 99, 114, 123
Crenulation cleavage 54, 104
Cretaceous System 38–39
Cristobalite 129, 168
Crushed rock aggregates 7, 65, 78, 157–158
Crystal Black dolerite 30, 33
Cwt-Y-Bugail slate 54, 102
Cyprus 38, 40

D-cracking 102
Dacite 28, 30, 172, 364–365
De-icing salts 112, 222
Decarbonation *116*
Decorative concrete 160
Delamination 58, 118–121
Delayed ettringite formation (DEF) 105–106, 210–211
Denmark 90
Dequesa slate 104
Devitrification 'stone' 170, 361
Diaquartzite 151
Dicalcium silicate 79, 169
Differential thermal analysis 76, 119
Digital technology 9, 13, 14
Dimension stone 7, 21
Discolouration, fire damage 49, 96, 115, 229
Dolerite 4–6, 29, 30, 33–35
Dolomite 41–42, *70*, 113, 116
    staining techniques 177
Dry shake topping 95, 191–192
Drying shrinkage 100, 197
Dunham limestone classification 34, *34*
Dyes 11, 76, 104

Earth construction 137, 163
Egypt 28–29, *28–30*, 41, 42, 68–69, 138
Electron microscopy 14, 76, 144
Encrustations 62, 125
Environmental scanning electron microscopy 14
Epidote 52
Epoxy resin 17, 18, 160, 328–329

Equatorial Guinea 36
Etchants 52
Ethiopia 27
Ettringite 83, 100, 102, *116*, 200, 203, 323
    delayed formation 105–106, 113, 210–211, 226
European standards 9, 22–23, *23*, 51
Excess voidage 92–93, 185

Faience 163, 166
Feldspars 20, 26, 29–30, *30*–34
    differentiation by staining 177
Ferrite phases 84, 169
Ferromagnesian minerals 25, 26, 29, 30–31, 38–40, 136
Fibre glass 142, **277**
Fibre-reinforced concrete 89, 121, **178**
Fibres 89, 121, **178**, 276–277, 326–327
Field emission scanning electron microscopy 14
Filetto Rosso limestone 47, 88
Filler joist concrete 62, 79, 128, 161
Fillers 89, 155, 171, 314, 366
Finely ground slices 18–20, 76, 92, 184, 191
Fines 68, 139
Finland 27
Fire damage
    concrete 114–117, *114*, *116*, 227–231
    masonry 49, 95–96
Flint aggregate 137, 143, 220, 222–223
Flint stone 43, **77**
Flintlime bricks 124, 126, 249–251
Floor finishes 129, 130, *130*
    screeds 130–132, 254–260
    synthetic resin 133–135, *134*, 264–266
    terrazo 132–133, 261–263
Floor slabs, concrete 95, 191–192
Flue gas desulfurization gypsum plaster 140–141, 271–273
Fluorescence microscopy 7–8, 11–12, 48, 76, 90, 179–182, 228
Fluorescent dyes 11, 17, 76
Fly ash 87, 174
Folk limestone classification 34, *35*
Fossil burrows 47, 89
Fossil remnants 24
    flint 43, **77**

limestone 34, 38, 40, 41, 46, 55, 58, 63, 69, 144, 283
    slate 54–55, 106
France 39, 138
Freeze-thaw mechanisms 48, 102
Fresco 152, **307**
Frost action/attack/damage 48, 91–92, 102, 151
Furuland Schist 45, *79*

Gabbro *22*, *25*, 29–30, 31, 157
Geomaterials 1, 7, 8
    economic value 8
Germany 64, 172
GGBS, *see* ground granulated blastfurnace slag
Gibbsite 119
Giza Pyramids 20, 26, 41, 68–69, 138
Glass *16*, 163, 170, 360–362
Glass fibres 89, 121, **277**
Glauconite 59, 160, 257, 309
Glue 17
Gneiss 44–45, *44*, 80
Grain, slate 54
Granite 20–23, 26–27
    heating *50*
    standards and quality *23*
    weathering 66
Granite aggregates 70–72, *70*, 112, 135, 146, 221
Granolithic screed 130, 132, 260
Graphite 52
Great Wall of China 148–150, 164, 298, 338–340
Greece 138
Greywacke aggregate 70, *70*, 110, 111, 144, 219
Grog 164, 166, 168, 351–354
Ground granulated blastfurnace slag (GGBS) 87–88, *87*, 175
Groundwater, sulfate attack 102, 203–204
Grouted macadam 172, 368–369
Gypsum 7, *8*, 105, *116*, 138, 142, 213, 258–259
    staining 178
Gypsum plaster 139–140, 267–275
Gypsum plate 6, 11
Gypsum-based plaster 138, 139–143, 267–278

HAC, *see* high-alumina cement and concrete
Hadrian's wall 144, 282
Haematite 52, 99, 139
Halite 112, 222

Hampole magnesian limestone 41, **70**
Hand specimen 10
Hard burnt lime 146, 148
Heating
    concrete 114–117
    masonry stone 48–50, *50*, 95–97
Hemihydrate 139, 140, 147, 267–268
High-alumina cement and concrete (HAC) 117–120, 232–237
Highmoor magnesian limestone 42, **72**
Hornblende 30
Huddlesdon magnesian limestone 42, **71**
Hydraulic lime 138, 143
Hydraulicity 138, 143
Hydrogen fluoride 84, 168–169
Hydrogen sulfide 108–109, 215

Igneous rocks 20–40, *22*, 24–31
    geological classification 25, *25*
Ilmenite 24
Image analysis 13, 90, 93, 98
Impala Black gabbro 31
Imperial porphyry 28–29, *28*
Impregnation 17, 104, 115
Inclusions 55–56, 110–113, 146–147, 148
India 45–46, 80, **87**
Indian Green marble 46, **87**
Infrared spectroscopy 13, 76, 89, 134, 144
Ironstone 42–43, 76
Italy 24, 40, 46, 47, 67, 83–84, 88, 173

Joints 32, 55
Jura limestone 40, 64
Jurassic limestones 35–36, 47–50, 144, 284

Ketton limestone 36, 47, 48
Killas 105, 209

Laitance 94, 123
Lapis Porphyrites 28, *28*
Lapis Porphyrites Niger 28–29, 30
Lapping machine 15, 18
Larvikite 26, 27
Leaching 100–101, 151, 199–200, 302
Leica Microsystems Ltd 2, 3
Lightweight aggregates 78, 159
Lignite 69, 109, 217

Lime (CaO) 86, 124, 137, 143
Lime products 143
Lime-based plaster/mortars/renders 138, 143–152
Limestone 7, 34–42
    classification *34, 35*
    deterioration and decay 48–49, 96
    standards and quality *23*
Limestone aggregates 78, 158
Limestone fillers 89, **177**
Limewash 152, 306–307
Limewater 69
Lincolnshire limestones 35–36
Logitech machines 15, 18
Lytag, sintered pulverised-fuel ash aggregate 62, **127**, 159

Macadam 7, 171–172, 364–365
Magnesian limestone 41–42, 70–72
Magnesite 52
Magnesium sulfate 106
Magnetite 52
Map cracking 106, 110, 113, 218, 226
Marble 7, 8, 45–46, 83–87, 94
    deterioration 48–49, 94
    heating 48–49, *50*
Marcasite 52, 56
Marine aggregates 62–63, 68, 77–78, 124–125, 284
Marl, metamorphosed 116–117
Masonry 7, 21, 137, 163
    deterioration 47–50, 90–95
    fire damage 49, 95–96
    *see also* Bricks
Masonry cement 155, 314
Mastic asphalt 129, 172
Materials engineer 8–9
Melilite **128**
Metakaolin *87*, 148
Metamorphic rocks *22*, 44–46, 79–87
Metaquartzite 45, 81, 152
Mica 68, 310
    in aggregates 68, 154, 310
Micrite 57–58, 64, 68–69
Microcracking 49, 95, 151
Microgranite *25*, 70, 147
Micrometric analysis 76
Microporosity 48, 151
Microporous aggregate 67, 137
Microscope 2–3, 10
Microscopy 10–12
Microsilica (silica fume) *87*, 88, 176
Migmatite *44*
Mineral additions 86–88

Mineral proportions estimation 19
Ming Dynasty 150, 164, 298
Mix proportions 150, 156, 160
Modal analysis of concrete 13, 93–94
Moh's scale of hardness 10, 77
Mokattam Formation 41, 68
Moleanos limestone 40, 66
Monks Park limestone 37, 52
Mons Porphyrites 28, *28*
Mortars 137–138
    binders 137–138, *137*
    gypsum 138, 152, 304
    lime 138, 143–152
    Portland cement-based 152–159
    repair 138, 159, 160, 161, 326–327
    specialist 9, 159–162
    workmanship 150–151, 156–158, 299–301, 320–321
Moss colonization 48, 93
Mudstone/siltstone 115, 119
Mullite 168
Mundic concrete 105, 208–209
Munsell colour charts 10
Muscovite mica 68, 208, 310, 353
Mylonite *44*, 150

Natural aggregate 61, 62
Natural cements 138
Nepheline syenite 27, 28
Nodules 55–56, 103
Norite 29–30, 32
Northowram sandstone 33, 41
Norway 26, 79, 81
Nubian sandstone 33, 42

Obsidian 24, *25*, 27
Oman 123
Onyx-marble 42, **75**
Ooliths 34, 36, 37, 47–50
Oolitic limestone 36, 47, 52–55
Opal 70, *70*
Ophicalcite 46, 85–86
Organic matter 68–69, 140–141, 217
Organic solvents 135
Orthoquartzite *32*, 72

Paints 152, 308
Particle size distribution 64
Paving 21, 23, 25
Penrhyn slate 99, 101
Peridotite 156
Perlite 141, **274**
Perryfield Roach limestone 37–38, 55

Petit Granit limestone 40, 63
Petrographic examination 8–12
Phenolphthalein 96
Photomicrography 9, 13–14
Phyllite 69, 209
Phyllosilicates 51, 52, 69, 98, 112
Piddington Roman Villa **287**, 293
Piemontite 29
Pigments (and colorants) 85, 122, 124, 132, 134, 156, 265–266, 318
Pit sand 144, 279–280
Plaster 137–138, 267–280
Plasterboard 139, 142, 278
Plastic cracking 99, 196
Plastic shrinkage 99, *99*
Plasticizers *87*, 89, 92, 152, 155, 156–158, 321
Pleochroism 118
Pleochroite 118, 234
Point-counting 24, 61, 76, 88, 92, 93, 150
Polarizing microscope 3, 8
Polished specimens 10, 12, 18–20
Polishing machine 18, 18
Pollution 47, 90
Polyphant stone 39
Polypropylene fibres 89, 160, 326–327
Pop-outs 109, 216–217
Popcorn carbonation 101, 254
Pore structure, stone 48, 151, 164
Porfido Rosso stone 25, 27
Porosity 48, 150, 164
Porous asphalt 172
Porphyroclasts 148–150
Porphyry 27
Portland cement 7, 79–89, 137, 153
    classification 79, *80–81*
    sulfate-resisting (SRPC) 83–84, 167, 170
    white 85, 156, 171, 315
Portland limestone 37–38, 52–54, 90, 92
Portlandite 86, 172, 199, 322
Portugal 40
Pozzolana 86–87, 148
Precast concrete 106, 121, 132
Production of geomaterials 8, *8*
Pulverised fuel ash (PFA) 62, 87–88, *87*, 124, 174, 326
Pumice 142, 148, 173, 292
Purbeck stone 38, 56–57
Pyramids, Giza 20, 26, 41, 68–69, 138
Pyrite 9, 10, 56, 69, 105, 109, 110–113, 216

Pyrrhotite 56–57

Quartz 23, 70, 146
Quartzite 72, 72, 81–82
Quicklime 124, 143, 146, 287

Rapakivi texture 23, 27
RapidAir 457 system 186
Recycled aggregates 64
Reduction spots 52, 101
Reed, chopped 142
Reflected light microscopy 10, 12
Reinforcement, concrete 75, 97–98, 123
Reinforcement corrosion 96, 124, 195
Render 137, 137, 138, 158, 324
Repair mortar 138, 159, 160, 161, 326–327
Resins 17, 17, 133–135, 134, 264–266
Rhyolite 25, 27
RILEM 150
Roach (Portland) stone 38, 55
Road aggregate 61, 66, 172–174, 363–369
Roman concrete and mortar 86, 138, 148–149, 281–282, 292–293
Roman Imperial porphyry 28
Roman mortar 138, 144, 281–282
Roman tile 148, 293
Roofing 51
Roofing slate 9, 10, 51
    classification 59, 60
    petrographic examination 12, 52
    properties 52–56, 98–112
    testing 51, 58–60, 59, 60, 121
    weathering and deterioration 56–58, 113–120
Rossa Verona limestone 40, 67
Royal Green dolerite 30, 35
Royal White granite 21, 26
Rubet porphyrites 28, 29
Rust staining 57, 98, 109, 114, 216
Rutile 52

Salt attack 102, 151, 308
Salt crystallization 47, 100, 166, 349–350
Sample preparation 17–20
Sampling 14–17
Sand 62, 66, 131–132

Sand:cement screed 130, 254–256
Sand carpet 172, 174, 367
Sand and gravel aggregates 61–62, 124–125
Sandlime brick 124, 246–248
Sandstone 11–12, 33–34, 41–44, 50, 91, 97
Saudi Arabia 69
Scanning electron microscopy (SEM) 13, 14, 76, 144
Schist 22, 44, 69, 79
Scoria 30, 37
Screeds 16, 130–132, 254–259
Sea-dredged aggregate 62–63, 124–125
Seawater attack of concrete 106–107, 199–203
Secondary deposits in concrete 100, 199–203
Sedimentary rocks 22, 32–43, 41–78
Segregation 94, 189
Septarian nodules 138
Serpentine 31, 39
Serpentinite 31, 38, 40
Sewers 108–109, 215
Sharp sand 66, 131, 144, 280
Shell 53, 56, 68, 124, 284
Shivakashi charnokite gneiss 45, 80
Sicily 37
Siderite 52
Silica fume, condensed (microsilica) 87, 88, 176
Silt 68
'Silver' sand 66, 134
Sky Blue marble 46, 85
Slag 87–88, 175
Slaked lime 143, 146
Slate 51, 69
    see also Roofing slate
Soft (building) sand 66, 132, 144, 279
South Africa 29
Spain 40, 65, 98, 104, 106
Spalling 98, 114–115, 124
Sparite 56
Staining techniques 20, 177–179
Stamford limestone 36, 50
Standards 9
Statuario Venarto marble 46, 83–84
Steel fibres 89, 178
Steel reinforcement 75, 97–98
Stone filler 159, 162, 337
Stone mastic asphalt (SMA) 172
Structural cracking 100, 198

Stucco 138
Stylolite 47, 67, 88
Sulfate attack 83, 102–104, 158–159, 203–204, 324–325
Sulfate-resisting Portland cement (SRPC) 84–85, 156, 167, 170
Sulfates 69, 102, 152
Sulfides 69, 108
Sulfuric acid 108, 215
Superplasticizer 89, 92
Surface deposits 47, 90, 166, 349–350
Surface finishes, masonry 152, 306–308
Sustainable construction 64
Swaledale fossil limestone 35, 46
Syenite 26, 27
Synthetic resin floor covering 129, 133–135, 264–266

Tanzania 63
Taynton limestone
Tension gashes 99, 150, 196, 301
Terracotta 7, 163, 166–168, 351–356
Terrazzo 129, 132–133, 261–263
Testing 8, 22–23, 23, 51
Tetracalcium aluminoferrite 84, 169
Thames Valley river gravel and sand 115, 144, 154, 279–280
Thaumasite 103, 104, 205–207, 325
Thaumasite form of sulfate attack (TSA) 17, 102–104, 158–159, 205, 324–325
Thermal analysis 76, 119
Thermal cycling 48–49, 94
Thin sections 10  11, 14, 16, 17–18
Thiobacillis 108
Tile adhesive 160–162, 330–336
Tile glazes 168, 357–358
Tiles, ceramic 162, 168–170, 337, 357–359
Totternhoe clunch 39, 59
Tourmaline 52, 134, 208
Travertine 42, 73–74
Trent Valley river gravel and sand 115, 155
Tricalcium aluminate 79, 169
Tricalcium silicate 79, 168
Tridymite 70, 70, 170, 355, 362
Triple blend cements 88
Tufa 202
Tuff 25, 25, 27, 226

Tura limestone 41, 69
Turkey 42

Ultracataclasite 149
Unburnt coal 124
United Arab Emirates 156
Unsoundness 66–69, 124, 135–141
Uranyl acetate 178

Veins (slate) 55, 57, 64, 107–109
Verde 31
Vermiculite 141, 275
Verona limestone 40, 67
Villar Del Rey slate 52, 98
Voids and voidage 92–93
Vtratza limestone 47, 89

Wadi gravel 62, 123, 156
Wallboard 139, 142, 278
Water:cement (W/C) ratio 89–92
Weathering 24, 56–58, 66–67, 100, 113–120, 135–136
Websites 13, 14
Westmorland Green slate 100
White (Portland) cement 85, 122, 132, 156, 171, 315–317
Workmanship 94–95, 150–151, 156–158, 189–192, 299–301, 320–321

X-ray diffraction analysis (XRD) 13, 164
X-ray fluorescence analysis (XRF) 13

Yellow Rock granite 22
York stone 33, 41

Zimbabwe Black dolerite 34